DEVELOPMENTS IN
PRESSURE VESSEL TECHNOLOGY—1

Flaw Analysis

DEVELOPMENTS IN PRESSURE VESSEL TECHNOLOGY—1

Flaw Analysis

Edited by

R. W. NICHOLS

D.MET., C.ENG., F.I.M., F.I.MECH.E., F.WELD.I.,
HON.F.BRIT.INST.NDT

Editor, The International Journal of Pressure Vessels and Piping
Chairman, Institution of Mechanical Engineers, Pressure Vessel Section
Chairman, International Council of Pressure Vessel Technology

APPLIED SCIENCE PUBLISHERS LTD
LONDON

APPLIED SCIENCE PUBLISHERS LTD
RIPPLE ROAD, BARKING, ESSEX, ENGLAND

British Library Cataloguing in Publication Data

Developments in pressure vessel technology.
1. : Flaw analysis
1. Pressure vessels
I. Nichols, Roy Woodward
681'.766   TS283

ISBN 0-85334-802-2

WITH 11 TABLES AND 81 ILLUSTRATIONS

Printed in Great Britain by Galliard (Printers) Ltd, Great Yarmouth

PREFACE

The current pace of technological development in many areas is such that technical books tend to become quickly dated. This situation has been recognised with respect to the volume *Pressure Vessel Engineering Technology* which I edited and Applied Science Publishers Ltd published in 1971, and we have sought some way of solving the problem. The result is to be a number of smaller books in a series to be called *Developments in Pressure Vessel Technology*, each of them devoted to a technical area roughly corresponding to that of a single chapter in the earlier book. The new developments of theory, practice and technique in each area will be discussed in turn in a format that is intended to be both convenient and practical. We have found that some compromise is needed on the contents of each chapter. On the one hand it should stand on its own, presenting in itself an overview of the current status of a particular area of technology. On the other hand, space limitations necessitate that much of the detailed background be only presented in summary so that as much as possible of the current volumes is concentrated on the more recent work. The authors and myself have tried to remember these aspects throughout the preparation of the books, which in each case takes the corresponding chapter of *Pressure Vessel Engineering Technology* as the starting point to which further reference should be made for more detail of the earlier background. Background material is therefore only repeated in these volumes in summary and in order to allow each chapter to be understood by itself.

A further advantage of the present approach is that it allows each subject area to be broken down into further detail, so that each specialist subject can be dealt with by an expert in that subject. In the selection of an author on each particular topic I have sought active workers in that particular field, ones who know what is going on internationally, and usually are themselves internationally known both for their work and for their clarity of presentation. Another aspect to which we have paid attention is speed of preparation, the aim being that each review will be prepared and published at least as rapidly as any technical journal paper, so that the review is as up to date as is practical. To some extent that has been facilitated by using authors who already know the work well; further to this and wherever possible, I have used authors from the UK. This is regrettable in that I would have liked to continue to emphasise the international flavour that is a feature of *The International Journal of Pressure Vessels and Piping*; but the gains resulting from direct and speedy communication between the various authors, the editor and the publisher were in my opinion sufficient to justify this approach on this occasion. The potential drawbacks were recognised, however, and each author was asked to ensure an international coverage in the contents of each chapter. Finally, it is recognised that these volumes will themselves eventually become out of date; when that time comes revised volumes will be prepared to fill the need. I would appreciate comments from anyone on content, authorship and changes that would help in such volumes.

To turn from the general aspects of the series to this particular volume. Chapter 4 by Cowan and Nichols in the earlier book was devoted to the prevention of fast fracture, and gave information on several methods of approach including that of applying linear elastic fracture mechanics (LEFM). Since the date of writing that chapter the use of LEFM in the design and assessment of pressure vessels has become more widespread, as indicated by the introduction of such methods into some of the sections and appendices of the *American Society of Mechanical Engineers (ASME) Boiler and Pressure Vessel Code*, particularly those relating to pressure vessels for light-water cooled nuclear reactors (Sections III and XI). A major development has been the use of such procedures as part of the assessment of whether it is satisfactory to operate a vessel when some sort of flaw has been detected and characterised, for example by nondestructive examination either before or at intervals between service. This present volume covers various developments aimed at providing a rational 'fitness-for-purpose' basis for acceptance/rejection decisions in such cases, the techniques of nondestructive examination being covered in the next volume of this series. After an overall review that provides the background and summaries of the many developments, the more important of these developments are described in more detail in the following chapters. The second chapter, for example, describes in detail the ASME Section III and Section XI approach based on LEFM while Chapter 3 gives the results of UK and International Institute of Welding (IIW) discussions for similar schemes to cover pressure vessels operating under more ductile conditions where an elasto–plastic fracture mechanics approach is more relevant. The next chapter goes even further in consideration of fracture conditions in which there would be appreciable ductility before fracture, detailing the Central Electricity Generating Board two-criteria approach involving the assessment of what flaw sizes would lead to plastic instability. It should be noted, however, that the emphasis in these chapters is directed specifically at the use of these various approaches in flaw assessment, and they do not give the full theoretical background, which is covered in more detail in the book edited by Professor Latzko, entitled *Post-Yield Fracture Mechanics*, published by Applied Science Publishers.

A common problem to all these techniques is that of assessing the material toughness, especially where the material shows slow stable crack growth before unstable fracture. These aspects are discussed in Chapter 5 by Chipperfield.

Having assessed the critical flaw size for a particular application, it is necessary to allow for growth in service, and to assist this part of the assessment the present status of fatigue crack growth information is reviewed. Another chapter covers the assessment of when crack arrest would occur, thus allowing consideration of situations where the high stress or low toughness conditions giving instability are only local.

It is important to recognise that there are uncertainties in all of the factors fed into the various calculations associated with flaw assessment, so that any such estimate should be a statement of probabilities rather than be a deterministic result. The recognition of this is important and fits in well with the present approaches to assessments of component reliability, and the book therefore closes with a chapter devoted to such probabilistics treatments. It is intended too that this chapter will give the reader insight into the areas of greatest uncertainty in fracture assessment.

Finally, I must thank the authors for their speedy contributions and all those too numerous to mention but who have helped by allowing us to quote their work or reproduce from their publications. As far as possible I have tried to give some uniformity to the nomenclature and units throughout the book and to avoid undue repetition between the chapters. I hope that readers will find this book informative, interesting and up to date in its approach.

CONTENTS

LIST OF CONTRIBUTORS

F. M. BURDEKIN

The University of Manchester Institute of Science and Technology, Department of Civil and Structural Engineering, P.O. Box 88, Manchester M60 1QD, UK.

C. G. CHIPPERFIELD

Risley Nuclear Power Development Laboratories, United Kingdom Atomic Energy Authority (Northern Division), Risley, Warrington WA3 6AT, Cheshire, UK.

A. COWAN

Risley Nuclear Power Development Laboratories, United Kingdom Atomic Energy Authority (Northern Division), Risley, Warrington WA3 6AT, Cheshire, UK.

B. J. L. DARLASTON

Research Division, Berkeley Nuclear Laboratories, Central Electricity Generating Board, Berkeley GL13 9DB, Gloucestershire, UK.

D. FRANCOIS

Université de Technologie de Compiègne, Département de Génie Mecanique, Centre de Recherche de Royallieu, Rue P. de Roberval, BP233/60206, Compiègne, France.

xi

J. D. HARRISON

　　The Welding Institute, Research| Laboratory, Abington Hall, Abington, Cambridge CB1 6AL, UK.

G. O. JOHNSTON

　　The Welding Institute, Research Laboratory, Abington Hall, Abington, Cambridge, CB1 6AL, UK.

I. L. MOGFORD

　　Research Division, Central Electricity Research Laboratories, Central Electricity Generating Board, Kelvin Avenue, Leatherhead KT22 7SE, Surrey, UK.

Chapter 1

INTRODUCTION

J. D. HARRISON

The Welding Institute, Research Laboratory, Abington, UK

1. THE DEVELOPMENT OF FLAW ACCEPTANCE STANDARDS

The recent safety record of the boiler and pressure vessel industry is excellent. Statistical surveys suggest that the risk of disruptive failure in service of a boiler drum made to current standards is less than 1 in 10^6 per drum or vessel year.[1] That this is so must be attributed in large part to the development of well founded codes and standards. It was not always the case. Slater[2] refers to the appalling history of boiler explosions in Mississippi river steam boats. Between 1816 and 1848, 1443 lives were lost as a result of these and the death toll reached its height in 1865 when no less than 1547 perished in a single disaster when the boilers of the steam boat *Sultana* burst.

These events and similar experiences in Europe led to public pressure for improved standards of construction. In many countries legislation was enacted in the second half of the nineteenth century regulating the use of boilers and requiring their inspection. This legislation gradually brought into being the forerunners of our present codes and standards.

The main intention of these codes is to ensure that the finished product conforms both in design and execution with normally accepted standards of construction. They have gradually evolved from the early relatively simple documents into present day codes, such as British Standard (BS) 5500[3] and the *American Society of Mechanical Engineers'* (*ASME*) *Boiler and Pressure Vessel Code*.[4,5,6] These cover every aspect of the product from design through material selection, fabrication, heat treatment, inspection, testing to in-service monitoring.

1

Although this book deals with the significance of flaws in general, it will usually be the case that these are associated with welds. Welding was introduced into the boiler and pressure vessel industry by the A. O. Smith Organisation in 1925.[7] It was appreciated at that time that the quality of the weld was very much in the hands of the individual operator. Welds were likely to be more heterogeneous than plate or forgings and, therefore, destructive tests on sample coupons might be less indicative of the overall quality of the welds in a structure than would similar tests be of the quality of wrought products. Furthermore, there was an inherent mistrust of this new joining process, welding, despite its obvious advantages over riveting. Thus an early concern to ensure the integrity of welds led to the introduction of radiography. It is probably no coincidence that the first work on X-ray inspection was completed by Union Carbide also in 1925. With the introduction of radiography it was necessary to fix acceptance levels for the defects which were found. In 1934 Lloyd's Register first published rules for welded pressure vessels with clauses dealing with radiographic inspection. The evolution of acceptance levels for weld defects in UK standards has been reviewed by Young.[8] The first standard giving detailed acceptance levels was issued by a group of insurance companies and appeared in the late 1930s; but only when BS 1500[9] was issued in 1958 were such acceptance criteria included in a British Standard. The acceptance criteria in reference 9 and those in many other national codes for boilers, pressure vessels and pipelines all seem to have a common parent, namely the ASME Codes.

The defect acceptance clauses in all these codes are coloured, not unnaturally, by the fact that they were written in terms of radiographic inspection. Having no scientific way of fixing acceptance levels with regard to the possible effects of defects on serviceability, the code writers worked in terms of what could be detected and readily measured by X-ray. Thus precise limits are given for the volumetric defects, slag and porosity whereas cracks, which are not so easily detected by radiography and certainly cannot be measured by it, are covered by bold statements such as, 'there shall be no cracks or crack-like defects'. The levels given for volumetric defects are essentially arbitrary or are based on the quality that might be expected of a 'good welder'. Even so the levels are stated with great precision, slag being limited by length and porosity by reference to a series of comparison charts. A story, probably apocryphal, has it that the latter were derived by the committee members sitting around the table at lunch time, making marks on paper napkins until they finally reached an agreed number of marks per square inch.

In 1952 the International Institute of Welding (IIW) published a set of reference radiographs. These have a useful function in assisting radiographers in the identification of the various types of defect observed. However, the mistake was made of grading these reference radiographs by a colour code ranging from black through green and brown to red for defects of increasing severity. Some Scandinavian standards use this colour coding to fix acceptance levels. It should be emphasised that this was never the intention of the IIW reference set and that this approach is as arbitrary as that of the ASME code.

Perhaps the most illogical requirement in some existing codes is the coupling of percentage radiograph (often only a small percentage of the total length of weld is inspected) with detailed, arbitrary, and rigorously enforced acceptance levels. By definition, percentage radiography can only be a quality control measure. If the safety of the structure depends on the elimination of defects found by radiography (in fact, as discussed later, this will usually not be the case) then 100 % inspection must be mandatory.

In recent years it has been appreciated that radiography is not reliably capable of detecting crack-like or planar flaws and certainly has little potential for sizing them. (It has been suggested that radiography is the nondestructive examination (NDE) technique best suited to the detection of harmless defects.) This has led to the increasing use of ultrasonic inspection. This was first permitted, as an alternative to radiography, in the 1965 revision to BS 1500,[9] provided that the technique was shown to be equally capable of detecting unacceptable defects (!)

2. THE CASE FOR CHANGE TO A 'FITNESS FOR PURPOSE' APPROACH

In the last 15 years, developments in ultrasonic techniques have been rapid and the method is now very widely used for critical components in Europe. In the USA its use at the construction stage is less widespread, but it has been accepted as the only practical technique for the in-service inspection of nuclear pressure vessels.

Unfortunately, developments in acceptance standards for new construction have not kept pace with those in NDE and such standards are still often couched in terms suitable only for radiographic inspection. The phrase, 'there shall be no cracks or crack-like defects', still appears, but becomes meaningless when applied to an NDE method which is already capable of detecting grain boundaries. The use of an NDE method of great sensitivity combined with arbitrary acceptance criteria has led to an increasing repair rate, although weld quality is almost certainly improving. The vast majority of this repair work is certainly quite unnecessary. Salter[10] reports a survey of the repairs actually carried out to the main seams of high quality pressure vessels made in the UK. About 84% of these were for slag inclusions, 3% for porosity and 13% for planar defects. It is only the latter which would be considered potentially deleterious from the point of view of serviceability. The economic consequences of this are considerable. Not only is money unnecessarily expended on the repairs, but there can be significant consequential losses in terms of delayed delivery and goodwill. A recent example of the economic consequences of arbitrary acceptance standards is provided by the Alyeska pipeline.[11,12] As a result of an audit of the radiographs carried out after the line had been laid, some 4000 defects which exceeded the levels allowed in the pipeline standard[13] were found. Repairs to almost all of these were carried out at a reputed cost of $52 m. One repair to a weld in a river crossing, which required the construction of a coffer-dam, is said to have cost 2\frac{1}{2}$ m. The repair itself took 3$\frac{1}{2}$ minutes and involved grinding off the cap run where the offending defects, a cluster of pores, were situated! An analysis showed that almost none of the 4000 defects could possibly impair the serviceability of the line and it is interesting that the principle of judging defects on a 'fitness for purpose' basis was accepted in this instance by the responsible US government agency in that the acceptance clause was waived in the case of 3 defects. These were at other river crossings and it is thus likely that the unnecessary expenditure of several further millions of dollars was saved by the belated application of more rational criteria.

Apart from the cost of arbitrary acceptance levels there could also be serious consequences in terms of safety. Repair welds are made under conditions of high restraint and there is a risk that a harmless, but readily detectable defect, such as a slag inclusion, will be replaced by a potential harmful crack, which is less easy to detect. One such case is cited by Lundin.[14] Here a number of repairs were made at the same location. The defects repaired are not mentioned, but since the inspection method was radiography, it is likely that they were slag inclusions. Finally, a crack developed across the weld. In this case it was large enough to be plainly visible; but there is no reason to believe that this would always be so.

3. THE ROLE OF NDE IN QUALITY CONTROL AND ACCEPTANCE

If it is accepted that irrational or 'good workmanship' standards are unsatisfactory, what is to take their place? It is suggested that the whole role of NDE in welding needs to be reconsidered. Should NDE be used primarily for making accept/reject decisions, or should it not rather be used for quality control? Harrison and Young[15] suggested that what was needed was a two-tier approach. Defect levels should be set for quality control purposes. These might be similar to existing acceptance levels. Defects smaller than these quality control limits would be automatically accepted, but larger defects would not be automatically repaired. The cause for the loss in quality would be examined and rectified. Repair would only be required if quality fell below an acceptance level based on 'fitness for purpose'. This approach would be similar to the quality control methods adopted almost universally outside the welding industry. The object of these is to ensure, by the use of quality control charts and by taking appropriate corrective action, that the rejection or repair rate is zero or very small.

Ultrasonic testing has great advantages over radiography for quality control, since it can be used to monitor work in progress. Radiography requires the complete interruption of the work and usually the removal of the vessel to a special bay. Ultrasonic testing can follow close after the completion of any given seam, so that action can be taken to rectify a procedure which is leading to a fall in quality.

A two-tier approach of the type outlined is already allowed for in the latest UK pressure vessel code, BS 5500, and in the pipeline standard, BS 4515. Both these permit the contracting parties to carry out an engineering assessment of defects which exceed the quality control levels without saying how this assessment should be done.

The ability to assess defects on a 'fitness for purpose' basis is useful, not only at the construction stage, but also for considering defects found by in-service inspection. Such assessments can be used to decide whether a repair is required and, if so, to determine the timing of the repair. This latter point has important economic implications in the power generation industry, where it is advantageous to delay repairs until a shut down period at a season when electricity consumption is low.

4. THE BASIS OF A 'FITNESS FOR PURPOSE' APPROACH

What is to be the basis for judging the accept/reject level in the two-tier approach? How is the significance of a given flaw to be assessed? It is perhaps not surprising that research into the effect of weld defects on performance began at virtually the same time as the introduction of radiography into codes. Homes[16] reported studies of the effect of weld defects on fatigue performance in 1938. Since then the literature on defect significance has burgeoned. Lundin[17] in a 1976 review refers to 322 publications and lists an additional 260 in an appended bibliography. The problem for the engineer is to distil, from this welter of information, that which is necessary to judge the flaws occurring in his particular structures.

An early attempt at such a distillation was made in 1968 by Harrison *et al.*[18] Following this publication, the British Standards Institution (BSI) established the WEE/37 Committee, whose remit was to produce rationally based weld defect acceptance criteria. After a prolonged gestation period, this Committee issued a Draft for Comment[19] in 1976. Further work on the draft is in hand and it is hoped that a document can be issued during 1979 (see Chapter 3).

Two important preliminaries to the engineering assessment of a defect are first to establish its type and then to consider the failure modes which might be influenced by its presence.

4.1. Defect Types
An IIW atlas[20] names a total of 83 different weld defects. It is convenient to divide these into 2 categories namely:

1.   Non-planar or volumetric defects (slag inclusions, porosity, etc.).
2.   Planar defects (cracks, lack of side wall fusion, etc.).

To these should be added a further category which has received little attention in the literature and is not as yet covered in the WEE/37 document, namely the departures from the design shape, misalignment and angular distortion.

4.2. Potential Failure Modes
The WEE/37 Committee identified the following potential failure modes:

1.   Brittle fracture.
2.   Fatigue.
3.   Yielding due to overload on remaining cross-section.
4.   Leakage in containment vessels.
5.   Corrosion, erosion, corrosion fatigue, stress corrosion.
6.   Instability (buckling).
7.   Creep/creep fatigue.

Although all of these must be examined in any complete assessment, it will usually be possible to limit detailed consideration to one or two.

A number of surveys have been carried out of the incidence of failure in boilers and pressure vessels (references 21–24). The Phillips and Warwick[21] and Smith and Warwick[22] surveys both show that operator error accounts for a significant proportion of all failures with low water operation of boilers figuring quite frequently. The number of failures by brittle fracture in service is very low. In the Phillips and Warwick and Smith and Warwick surveys this mode accounted for only 0·2 % of service failures. These were to associated pipe work and headers rather than to pressure vessels as such. Mosio[25] describes one of the few recorded service brittle fractures of a pressure vessel. Brittle fractures do occur, however, more frequently during the pre-service hydrostatic test.[26–29] An important effect of the hydrostatic test is to remove from the total pressure vessel population a number which would have been of marginal quality. Lidiard[30] shows that this has a marked effect in increasing subsequent reliability (see Chapter 8). Apart from this, the hydrostatic test has beneficial effects on the vessels that do survive it by relieving tensile residual stresses, blunting cracks and introducing compressive residual stress at their tips.[31]

In spite of the low incidence of service failure by brittle fracture, considerable effort has been expended on research into methods of specifying steels to avoid it and rightly so. Firstly, we obviously wish to reduce the incidence of hydrostatic test failures which are in themselves dangerous and expensive. Secondly, brittle fracture in service can be highly disruptive leading to large releases of stored energy, whereas other types of failure may well be successfully contained. This is of particular importance in the context of light water reactors (see Chapter 3) where failure of the pressure vessel cannot be covered by any engineered safeguard.

Some of the failures described by Smith and Hamilton[27] and those of references 25 and 29, indicate that if the material or heat treatment conditions have been incorrectly chosen, the size of defect required to initiate a failure becomes extremely small, far too small for confident detection by any known NDE method. On the other hand, the boiler drum failure described by Ham[28] showed how very tolerant to defects a ductile material can be. This drum failed from a 90 mm deep × 340 mm long defect; but even then, only after several pressurisations to full hydrostatic pressure. The defect is believed to have occurred as a result of stress relief. Even the most rudimentary inspection should have found this defect had it been carried out after this heat treatment. The message should be that brittle fracture is a problem in material selection.

If the material is susceptible the problem will not be overcome by refined NDE, whilst if the material is ductile refined NDE will not be necessary.

The survey by Collins and Monack[24] shows that in chemical plant one form of corrosion or another accounts for no less than 50 % of all failures. It may, therefore, be found surprising that these failure modes are given no further consideration here and are only dealt with briefly in reference 19. This is because corrosion is not usually associated with weld defects, as such, but rather with the use of materials that are incompatible with the environment.

Fatigue is an important cause of failure in welded plant. The survey by Phillips and Warwick[21] indicated that 26 % of pressure vessel failures could be attributed to this failure mode. Collins and Monach[24] report that it accounted for about 7 % of failures in piping and equipment. However, the conclusion drawn from these surveys is that fatigue is primarily a design problem. This conclusion is supported by Lancaster,[23] by an IIW survey of service fatigue failures[32] and by Welding Institute experience. The Institute is involved in investigating a number of fatigue failures every year, but the writer cannot recall one associated with a weld defect in the normally accepted sense of the term. Many fatigue failures initiate at the toes of fillet welds. In these regions there are small slag intrusions;[33] but these are present at all weld toes.[34]

In the following section there are brief introductions to the assessment methods that are available for brittle fracture and fatigue. These will be covered in greater detail in later chapters.

5.   BRIEF INTRODUCTION TO LINEAR ELASTIC FRACTURE MECHANICS (LEFM)

The major development which has made the rational assessment of flaws possible, particularly for brittle fracture, but also for other failure modes, was the introduction of modern Fracture Mechanics. The basis for this was established in 1920 by Griffith[35] who expressed the conditions for fracture propagation in energetic terms. However, widespread engineering application had to wait until 1957 when Irwin[36] showed that there was an equivalence between the energetic relationships of Griffith and a theory of fracture based on stress analysis of the near crack tip region. Irwin first analysed the stresses in an infinite elastic sheet with a through crack, length $2a$. The sheet was subjected to a uniform remote stress, σ, in a direction perpendicular to the crack plane. At a point defined by the polar co-ordinates r, θ with a crack tip as origin the stresses are:

$$
\begin{bmatrix} \sigma_{xx} \\\\ \sigma_{yy} \\\\ z_{xy} \end{bmatrix} = \frac{\sigma\sqrt{\pi a}}{\sqrt{2\pi r}} \begin{bmatrix} \cos\frac{\theta}{2}\left(1 - \sin\frac{\theta}{2}\sin\frac{3\theta}{2}\cdots\right) & \text{(a)} \\\\ \cos\frac{\theta}{2}\left(1 + \sin\frac{\theta}{2}\sin\frac{3\theta}{2}\cdots\right) & \text{(b)} \\\\ \cos\frac{\theta}{2}\left(\quad\sin\frac{\theta}{2}\cos\frac{3\theta}{2}\cdots\right) & \text{(c)} \end{bmatrix} \qquad (1)
$$

where σ_{xx} and σ_{yy} are the stresses in directions parallel to and perpendicular to the crack plane respectively and z_{xy} is the shear stress referred to this orientation. In eqns. (1(a)–(c)) the term $\sigma\sqrt{\pi a}$ is common and independent of r and θ. This is known as the stress intensity factor K_I (the subscript I refers to the opening mode of loading which has the greatest relevance to practical applications). Thus K_I is a single parameter defining the whole elastic stress field around the crack tip. As implied for the simple case of the uniformly stressed infinite sheet:

$$K_I = \sigma\sqrt{\pi a} \qquad (2)$$

A more general expression for the stress intensity factor is:

$$K_I = Y\sigma\sqrt{\pi a} \qquad (3)$$

where Y is a factor which takes account of other more complex geometries. Solutions for Y exist for many practical situations, see for example references 37 and 38.

6. APPLICATION OF LEFM TO BRITTLE FRACTURE ASSESSMENT

If K_I describes the elastic stress field, it would be reasonable to expect that for elastic materials there would be a critical value at which fracture would initiate. This proves to be the case. The critical value, K_{Ic}, is defined as the plane strain fracture toughness.

Standardised methods for measuring K_{Ic} were developed by the American Society for Testing and Materials (ASTM) E.24 Committee[39] whose efforts were paralleled by similar work in the UK.[40] The test is carried out on specimens of various possible geometries all containing notches which have been sharpened by fatigue cracking so that the conditions in a cracked structure may be simulated. The displacement of a clip gauge mounted across the open end of the notch is recorded against load. This is used to confirm that the specimen has behaved elastically. The K_{Ic} is calculated from the load at failure. When expressed in terms of the critical conditions, the usefulness of eqn. (3) in flaw assessment is obvious, allowing, as it does, the critical flaw size to be related to the applied stress and material toughness.

The above is the relatively simple and now almost universally accepted basis for linear elastic fracture mechanics (LEFM).

The method is strictly only applicable to materials which fail in a relatively brittle or elastic manner and for this reason there are limitations to its direct usefulness (see below). Nevertheless, LEFM forms the basis for material selection in a non-mandatory Appendix to ASME III[4] and for flaw assessment in ASME IX[6] (see Chapter 2). It is also the basis used, for situations where the total stress is below yield, in the WEE/37 document[19] (see Chapter 3).

7. YIELDING FRACTURE MECHANICS

As stated, eqn. (1) applies only for elastic behaviour. It will be seen in this equation that the stresses tend to infinity as the crack tip is approached ($r \to 0$). In engineering materials this is prevented by yielding at the crack tip. As long as the yielded region is small compared to the other dimensions of the body, eqn. (1) is still a good description of the stress field. However, as the yielded zone grows, it becomes increasingly inaccurate. To cater for this, much of the ASTM Committee's efforts were devoted to defining limiting size requirements set to ensure that the plastic zone in a K_{Ic} test is small compared to the specimen dimensions.

In practice it is only in relatively brittle materials that these validity limits can be met in section thicknesses of interest. If the structure is not thick enough for full thickness specimens to meet the validity requirements, recourse must be made to yielding fracture mechanics. The methods used are less mathematically rigorous and none are universally accepted. In the UK most effort has been devoted to the crack opening displacement (COD). The test method was issued by the BSI as a Draft for Development in 1972. This was revised and issued as a full standard in 1978.[41]

Test pieces and instrumentation are identical to those for K_{Ic} testing. The COD at the crack tip is calculated from two components. The elastic component is based on LEFM and is calculated from the load at failure. The plastic component is calculated from the opening displacement at the crack mouth measured with a clip gauge using a formula based on theoretical and observed rotational behaviour of the bend specimen as it collapses plastically. The basis for the calculation method which appears in reference 41 and which differs somewhat from that in the Draft for Development has been given by Dawes.[42] The test is carried out on specimens of the full thickness of the structure of interest. The fact that the test method is fully compatible with that for K_{1c} makes it possible for the test result to be interpreted as appropriate; K_{1c} if the specimen behaves elastically; COD if the validity requirements for K_{1c} are not met. The most widely adopted method of application of COD is the so-called design curve.[43] This expresses the relationship between fracture toughness, COD, defect size and applied strain. The design curve forms the basis of the yielding fracture mechanics approach in the WEE/37 document[19] (see Chapter 3). The design curve is largely empirical, being based on an analysis of a considerable number of large-scale test results where COD values obtained from small specimens were also available. The application of the design curve in many practical situations was reviewed by Harrison et al.[44]

The design curve takes the form:

$$\frac{\delta}{2\pi e_y \bar{a}} = \left(\frac{e}{e_y}\right)^2 \qquad \text{for } \frac{e}{e_y} \leq 0.5 \qquad (4(a))$$

$$\frac{\delta}{2\pi e_y \bar{a}} = \frac{e}{e_y} - 0.25 \qquad \text{for } \frac{e}{e_y} \geq 0.5 \qquad (4(b))$$

where δ = COD; e_y = yield strain = σ_y/E; e = total applied strain; and a = half length of through thickness crack. Dawes and Kamath[45] show that this curve represents about 95% confidence in survival.

Recently, mainly as a result of success with the application of the WEE/37 method to the Alyeska pipeline and to other oil industry problems, there has been an awakening of interest in COD in the USA. However, until this recent interest, the main US activity has been devoted to Rice's J contour integral.[46] J, which is only strictly meaningful for non-linear elasticity, is a path independent work integral around the crack tip. The emphasis of the US work has been the development of a test method which would give a critical value of J, J_{1c}, from a test on a small specimen which would behave elasto-plastically.[47] It is claimed that this value of J_{1c} can be used to predict the elastic value of K_{1c} that would be obtained from a much bigger specimen meeting the validity requirements for K_{1c}. This approach has only been shown to work for materials where the micro-mode of failure is the same (micro-void coalescence) for both small and large specimens. Workers in the UK have strong reservations concerning the general validity of the method. Dawes[42] shows that it is possible for the large specimen to fail by cleavage whilst the smaller one fails by micro-void coalescence. When this happens the value of K_{Ic} calculated from J_{Ic} for the small specimens may be much greater than the valid K_{Ic} for cleavage for the larger specimens. However, this is only a criticism of the particular direction taken by US developments. There are broad similarities between J and COD as outlined in Dawes' review.[42] Both are parameters characterising conditions at the crack tip. Both can be calculated from load displacement records on notched bend test pieces. Rice[46] shows that:

$$J = m\sigma_y\delta \tag{5}$$

where m is a plastic constraint factor, σ_y the yield stress and δ, the COD.

A J design curve was proposed by Begley et al.;[48] but there seems to have been little effort devoted, in the USA, towards furthering this idea. Turner[49] has recently re-examined the concept of a J design curve. He suggests either a series of curves dealing with specific geometric situations or a single blanket curve which lies on the conservative side of all of these.

Harrison et al.[44] and Merkle[50] show the close relationship that exists between the COD and J design curves based on eqns. (4) and (5). It can be shown that the COD design curve coincides with the upper and lower limit J design curves of Turner[49] if $m = 2$ and 1 respectively.

For ductile materials with high values of COD, the design curve will indicate that large defects can be tolerated. If the defects under consideration are part wall surface or buried defects, the WEE/37 document[19] requires that a check be carried out to ensure that plastic collapse of the remaining ligaments does not occur. The requirement is that the stress on the remaining ligament should not exceed the flow stress defined as $(\sigma_y + \sigma_u)/2$ where $\sigma_y = $ yield stress and $\sigma_u = $ tensile strength. Thus the WEE/37 document might be described as presenting a three-criteria approach to flaw assessment:

K_{Ic} for low toughness.
COD for intermediate toughness.
Flow stress for fully ductile.

Workers at the UK Central Electricity Generating Board (CEGB)[51] have put forward a two-criteria approach. For the two extremes of low toughness and full ductility the bases are as above. The intermediate range is catered for with a smoothing function based on the COD. This approach is described in Chapter 4.

8. ELONGATED PARTIAL WALL DEFECTS

Rigorous elastic–plastic solutions are only available for the simplest geometries, but LEFM can give a guide to the effect of various geometric parameters. Of particular importance in considering weld defects is the effect of elongated buried and surface defects. Irwin[52] first presented stress intensity factor solutions for the ends of the minor and major axes of buried elliptical or surface semi-elliptical defects. These have since been revised, but the main point that emerges is that, once the ratio of the length to depth of a defect exceeds about 5, the length becomes relatively unimportant and defect depth has an overriding effect. There are two important corollaries. Firstly, NDE methods must be developed to measure defect depth. Radiography has little potential here. Secondly, slag inclusions which are inherently long shallow defects are likely to have little effect. This is borne out in a practical way by large-scale tests reported by Ishii *et al.*,[53] who found that slag inclusions only affected fracture resistance at exceptionally low temperatures when their effect was no more harmful than that of the simple weld profile. Similarly porosity, firstly because individual pores are usually relatively small and secondly because of its rounded nature, has little effect on fracture resistance.[54]

9. FATIGUE ASSESSMENT

In spite of the fact that weld defects as such are seldom the initiation of fatigue failure, it is important to take account of this failure mode in any flaw assessment. Much research has been devoted, over the years, to the development of methods to make this possible.

10. PLANAR DEFECTS

Just as the stress intensity factor can be used to describe the crack tip conditions at brittle fracture initiation, so can the range in stress intensity factor be related to fatigue crack growth rates. This relationship, proposed by Paris and Erdogan,[55] is normally expressed as:

$$\frac{da}{dN} = C \Delta K^m \qquad (6)$$

where da/dN = crack growth rate; ΔK = range of stress intensity factor; and C and m = material constants.

In fact, plots of $\log da/dN$ versus $\log \Delta K$ are usually observed to be sigmoidal in shape. At low values of ΔK, da/dN drops rapidly and there is for some materials and environments a threshold value of ΔK (e.g. reference 56) ΔK_0, below which no propagation occurs. At high values of ΔK, da/dN accelerates rapidly as the peak value of K approaches the critical value for static failure.

Most crack propagation studies concentrate on the central linear portion where eqn. (6) applies. Studies of crack growth in a range of weldments and steels have been made by Maddox[57] and by Richards and Lindley[58] *inter alia*.

Environment and cyclic frequency can dramatically affect crack growth rates. Vosikovsky[59] shows that growth rates accelerate under cathodic charging conditions in a hydrogen sulphide environment at slow cyclic frequencies. Similar effects are reported by Kondo *et al.*[60] for slow frequency tests in a boiling water reactor environment. The rates assumed in Section XI of the *ASME Boiler and Pressure Vessel Code*, take account of this effect (see Chapter 2). On the other hand, James[61] finds that the growth rate in stainless steel in high temperature liquid sodium is reduced below the rate in air and is similar to that in vacuum. The reason for quoting this work is to show that environment can either accelerate or decelerate crack growth rates compared to those in air and that these effects on C and m in eqn. (6) must be taken into account. However, an observation which is important when one is assessing buried weld defects is that, below the creep range in pressure vessel steels, temperature, as such, has no noticeable effect on growth rate.[62]

Flaw assessment using eqn. (6) is achieved by integration from the initial flaw size to the final size at which leakage occurs or at which another failure mode intervenes. Harrison[63] has shown that this method can be applied to predict the results of a large number of fatigue tests on butt welds containing lack of penetration defects.

11.   NON-PLANAR DEFECTS

Because non-planar defects are such a common cause of repair and because there will, for such defects, be a significant number of cycles prior to initiation, it is worth treating them differently from the crack-like defects.

The literature on the fatigue strength of welds containing porosity and slag is considerable.[54,64] The results indicate that buried porosity at any level likely to occur in practice is only significant at exceptionally high fatigue stresses/lives, too high for all practical applications. Thus buried porosity need only be limited inasmuch as it interferes with the detection of other potentially more harmful defects.

Harrison and Doherty[65] have re-analysed the data on the effect of slag inclusions on a statistical basis. It is now possible to state the length of inclusion that can be allowed to give $97\frac{1}{2}$ or $99\frac{1}{2}$ % probabilities of survival at a given stress/endurance. The stress/endurance values chosen at which to calculate the allowable lengths are those appropriate to the various joint classifications in the fatigue clause of BS 5400, the bridge code. This work forms the basis for the WEE/37 approach to fatigue of welds with non-planar defects[19] (see also Chapter 3). Harrison et al.[62] report pulsating pressure fatigue tests on cylindrical vessels with defective longitudinal seams and on spherical vessels with defective nozzle welds. These have shown that the results described above of tests on small specimens can be applied with confidence to actual vessels.

12. A NOTE ON ALLOWABLE DEFECTS IN VESSELS DESIGNED TO ASME III FATIGUE DESIGN CURVES

Section III of the ASME code[4] has a series of fatigue design curves for different materials. These are $S-N$ curves based, it is understood, on strain controlled low cycle fatigue tests on plain materials with a limited number of additional tests on weld metals.[66] Both Harrison[67] and Jerram[68] have drawn attention to the very high design stresses allowed by these $S-N$ curves in terms of the fatigue behaviour of defect free joints. Recent work at the Welding Institute[62] draws attention to the same point with regard to the significance of weld defects. In the tests on model vessels of reference 62 a number of specimens failed at stresses/endurances lying below the ASME III fatigue design curves. Some of these failures were from small defects in machined butt welds. The defects causing failure would be acceptable even to existing acceptance criteria. The remedy, in the present writer's opinion, is to amend the fatigue design rules and not to tighten up on inspection levels. As they stand the $S-N$ curves imply a high probability of failure from ordinary design details.

BS 5500[3] also contains a fatigue design clause, but the $S-N$ curve in this lies at stresses which are a factor of about 1·6 lower than those in ASME III (alternatively the endurances in BS 5500 are a factor of 6 lower than those in ASME III). The work of Harrison[67] and Jerram[68] and the model pressure vessel tests on vessels with defective seams of Harrison et al.[62] all suggest that the BS 5500 fatigue design curve is conservative.

13. CONCLUDING REMARKS

This chapter has introduced briefly the topics which will be dealt with more fully later. To summarise:

1. There are good economic and safety reasons for changing from existing acceptance criteria for weld defects in pressure vessels.

2. A two-tier system is proposed which uses NDE as a quality control tool. Repair of defects is a last resort only to be adopted where the serviceability of the vessel could be impaired.

3. A 'fitness for purpose' approach must take cognisance of all possible failure modes.

4. Of these probably the most common in pressure vessels is corrosion; but, because this is usually a question of selecting materials for the environment, and not one of the significance of defects, it has not been considered further here.

5. Discussion here has been confined to brittle fracture and fatigue. The former is seen as a problem in material selection and the latter as one of design. If both these are right, structures will have a high tolerance to defects.

6. Fracture mechanics is a satisfactory method for assessing defect significance for brittle fracture. This can be based on LEFM and K_{Ic} for relatively brittle materials, on yielding fracture mechanics and COD for more ductile materials and on collapse for full ductility.

7. Fracture mechanics can also be used to assess planar defects under fatigue loading. Non-planar defects are relatively innocuous with regard to fracture, their effect on fatigue can be assessed by reference to $S–N$ diagrams for various defect severities.

8. ASME III fatigue curves are set at such a high level that very small defects within current code allowable levels could cause premature failure.

REFERENCES

1. BUSH, S. H. *Trans. ASME, J. Pressure Vessel Technology*, 1975, **97J**, 54.
2. SLATER, D. Standards and the non-ferrous vessel fabricator. *Proc. Conf. on Pressure Vessel Standards, The Impact of Change*, 1972, The Welding Inst., Cambridge.
3. *BS* 5500:1976, 'Unified fusion welded pressure vessels'. British Standards Inst. (BSI), London.
4. *ASME Boiler and Pressure Vessel Code*, Section III, 'Nuclear power plant components', 1977, ASME, New York.
5. *Ibid*, Section VIII, 'Pressure vessels'.
6. *Ibid*, Section XI, 'Rules for in-service inspection of nuclear power plant components'.
7. HOULDCROFT, P. T. *Metal Construction*, 1973, **5**, 443.
8. YOUNG, J. G. *Brit. J. NDT*, 1976, **18**, 149.
9. *BS* 1500:1958, 'Fusion welded pressure vessels for use in the Chemical, Petroleum and Allied Industries', British Standards Inst., London.
10. SALTER, G. R. and GETHIN, J. W. An analysis of defects in pressure vessel main seams. *Proc. Conf. on Pressure Vessel Standards, The Impact of Change*, The Welding Inst., Cambridge.
11. HARRISON, J. D. *Welding Inst. Res. Bull.*, 1977, **18**, 93.
12. MCHENRY, H. I., READ, D. T. and BEGLEY, J. A. 'A fracture mechanics evaluation of the significance of defects in pipeline girth welds'. Paper presented at ASTM Symposium, Elastic–Plastic Fracture, Nov. 1977, Atlanta.
13. *API* 1104, 'Standard for welding pipelines and related facilities', 1977, American Petroleum Inst., Washington, DC.
14. LUNDIN, S. Some examples of the consequences of common defects in welds. *Proc. 2nd Conf. on Significance of Defects in Welds*, 1969, The Welding Inst., Cambridge.
15. HARRISON, J. D. and YOUNG, J. G. *J. Roy. Inst. Naval Architects*, April 1975, 95.
16. HOMES, G. A. *Arcos*, 1938, **15**, 1951.
17. LUNDIN, C. D. *WRC Bull.* No. 222, 1976, Welding Res. Council, New York.
18. HARRISON, J. D., BURDEKIN, F. M. and YOUNG, J. G. A proposed acceptance standard for weld defects based upon suitability for service. *Proc. 2nd Conf. on Significance of Defects in Welds*, 1969, The Welding Inst., Cambridge.
19. *British Standard Draft for Public Comment* 75/77081, D.C., 1976, The British Standards Inst., London.
20. ANON. *Welding in the World*, 1969, **7**, 200.
21. PHILLIPS, C. A. G. and WARWICK, R. G. A survey of defects in pressure vessels. UKAEA Report AHSB(S)R 162, 1968, HMSO, London.
22. SMITH, T. A. and WARWICK, R. G. *Int. J. of Pressure Vessels and Piping*, 1974, **2**, 283.
23. LANCASTER, J. F. *Petroleum Int.*, 1974, **14**, 38.
24. COLLINS, J. A. and MONACK, M. L. *Materials Protection and Performance*, 1973, **12**, 11.
25. MOSIO, T. *Metal Construction*, 1972, **4**, 3.

26. BANKS, B. *Welding and Metal Fab.*, 1973, **41**, 4.
27. SMITH, N. and Hamilton, I. G. *J. West of Scotland Iron and Steel Inst.*, 1968/69, **76**, 111.
28. HAM, W. M., *Discussion to Conf. on Practical Applications of Fracture Mechanics to Pressure Vessel Technology*, 1971, Inst. Mech. Engineers, London.
29. ANON. *BWRA Bull.*, 1966, **7**, 149.
30. LIDIARD, A. B. and WILLIAMS, M. *J. Brit. Nucl. Energy Soc.*, 1977, **16**, 207.
31. NICHOLS, R. W. *Brit. Weld. J.*, 1968, **15**, 21 and 75.
32. ANON. 'Fatigue fractures in welded constructions', 1967, Publications de la Soudure Autogène, Paris.
33. SIGNES, E. G., BAKER, R. G., HARRISON, J. D. and BURDEKIN, F. M. *Brit. Weld. J.*, 1967, **14**, 108.
34. WATKINSON, F., BODGER, P. H. and HARRISON, J. D. The fatigue strength of welded joints in high strength steels and methods for its improvement. *Proc. Conf. on Fatigue of Welded Structures*, 1970, The Welding Inst., Cambridge.
35. GRIFFITH, A. A. *Phil. Trans. Roy. Soc. Series A*, 1920, **221**, 163.
36. IRWIN, G. R. *Trans. ASME, J. Appl. Mech.*, 1957, **24**, 361.
37. ROOKE, D. P. and Cartwright, D. J. *Compendium of Stress Intensity Factors*, 1976, HMSO, London.
38. TADA, H., PARIS, P. and IRWIN, G. *The Stress Analysis of Cracks Handbook*, 1973, Del Research Corp., Hellertown, Pa.
39. *ASTM E*399-74, 'Standard test method for plane strain fracture toughness of metallic materials', 1977, ASTM, Philadelphia.
40. *BS* 5447:1977, 'Methods of test for plane strain fracture toughness (K_{Ic}) of metallic materials', British Standards Inst., London.
41. 'Methods for crack opening displacement (COD) testing', 1978, British Standards Inst., London.
42. DAWES, M. G. 'Elastic-Plastic Fracture Toughness based on the COD and *J* contour integral concepts'. Paper presented at ASTM Symposium, Elastic–Plastic Fracture, Nov. 1977, Atlanta.
43. BURDEKIN, F. M. and DAWES, M. G. Practical use of linear elastic and yielding fracture mechanics with particular reference to pressure vessels. *Proc. Conf. on Practical Applications of Fracture Mechanics to Pressure Vessel Technology*, 1971, Inst. Mech. Engineers, London.
44. HARRISON, J. D., DAWES, M. G., ARCHER, G. L. and KAMATH, M. S. 'The COD approach and its application to welded structures'. Paper presented at ASTM Symposium, Elastic–Plastic Fracture, Nov. 1977, Atlanta.
45. DAWES, M. G. and KAMATH, M. S. 'The crack opening displacement (COD) design curve approach to crack tolerance'. Paper to be presented at Inst. Mech. Engineers Conf. on Tolerance of Flaws in Pressurised Components, 1978, London.
46. RICE, J. R. *Trans. ASME, J. Appl. Mech.*, 1968, **35 E** 379.
47. LANDES, J. D. and BEGLEY, J. A. 'Test results from *J*-integral studies: an attempt to establish J_{Ic} testing procedures'. *ASTM STP* 560, 1974.
48. BEGLEY, J. A., LANDES, J. D. and Wilson, W. K. 'An estimation model for the application of the *J* integral'. *ASTM STP* 560, 1974.
49. TURNER, C. E. An analysis of the fracture implications of some elastic–plastic

finite element studies. *Proc. Conf. on Numerical Methods in Fracture Mechanics*, 1978, Swansea Univ.

50. MERKLE, J. G. *Int. J. of Pressure Vessels and Piping*, 1976, **4**, 197.
51. DOWLING, A. R. and TOWNLEY, C. H. A. *Int. J. of Pressure Vessels and Piping*, 1975, **3**, 77.
52. IRWIN, G. R. *Trans. ASME, J. Appl. Mech.*, 1962, **29E**, 651.
53. ISHII, Y., KIHARA, H. and TADA, Y. *J. of NDT (Japan)*, 1967, **16**, 319.
54. HARRISON, J. D. *Metal Construction*, 1972, **4**, 99.
55. PARIS, P. C. and ERDOGAN, F. *Trans. ASME, J. Bas. Eng.*, 1963, **85 D**, 528.
56. FROST, N. E. and GREENAN, A. F. *J. Mech. Eng. Sci.*, 1967, **9**, 234.
57. MADDOX, S. J. *Welding Res. Int.*, 1974, **4**, 36.
58. RICHARDS, C. E. and LINDLEY, T. C. *J. Eng. Fracture Mechanics*, 1972, **4**, 951.
59. VOSIKOVSKY, O. *Closed Loop*, 1976, **6**, 2.
60. KONDO, T., KIKUYAMA, T., MAKAJIMA, H. and SHINDO, M. 'Fatigue crack propagation behaviour of ASTM A533B and A302B steels in high temperature aqueous environment'. Paper 6 HSST programme 6th Annual Info. Meeting, 1972, ORNL.
61. JAMES, L. A. *Atomic Energy Review*, 1976, **14**, 37.
62. HARRISON, J. D., ARCHER, G. L. and BOULTON, C. F. 'Significance of weld defects in pressure vessels under fatigue loading'. Paper to be presented at Inst. Mech. Engineers Conf. on Tolerance of Flaws in Pressurised Components, 1978, London.
63. HARRISON, J. D. 'The analysis of fatigue test results for butt welds with lack of penetration defects using a fracture mechanics approach'. Paper in *Fracture 69*, 1969, Chapman and Hall, London.
64. HARRISON, J. D. *Metal Construction*, 1972, **4**, 262.
65. HARRISON, J. D. and DOHERTY, J. A re-analysis of fatigue data for butt welded specimens containing slag inclusions. *Welding Res. Int.*, 1978, **8**, 81.
66. TAVERNELLI, J. F. and COFFIN, L. F. *J. Am. Soc. Metals, Trans.*, 1959, **51**, 438.
67. HARRISON, J. D. Low cycle fatigue tests on welded joints in high strength steels. *Proc. Conf. on Fatigue of Welded Structures*, 1970, The Welding Inst., Cambridge.
68. JERRAM, K. An assessment of the fatigue behaviour of welded pressure vessels. *Proc. 1st Conf. on Pressure Vessel Tech.*, 1969, ASME, New York.

Chapter 2

THE APPROACH TO ANALYSIS OF SIGNIFICANCE OF FLAWS IN ASME SECTION III AND SECTION XI

A. Cowan

UKAEA, Risley Nuclear Power Development Laboratories, Risley, UK

SUMMARY

ASME III Appendix G and ASME XI Appendix A describe linear elastic fracture mechanics methods to assess the significance of defects in thick-walled pressure vessels for nuclear reactor systems. The assessment of fracture toughness, K_{Ic}, is based upon recommendations made by a Task Group of the USA Pressure Vessel Research Committee and is dependent upon correlations with drop weight and Charpy V-notch data to give a lower bound of fracture toughness K_{IR}. The methods used in the ASME Appendices are outlined noting that, whereas ASME III Appendix G defines a procedure for obtaining allowable pressure vessel loadings for normal service in the presence of a defect, ASME XI Appendix A defines methods for assessing the significance of defects (found by volumetric inspection) under normal and emergency and faulted conditions. The methods of analysis are discussed with respect to material properties, flaw characterisation, stress analysis and recommended safety factors; a short discussion is given on the applicability of the data and methods to other materials and non-nuclear structures.

NOMENCLATURE

B thickness of test piece
C_0 constant used in Paris' equation for crack growth

K_I stress intensity factor

K_{Ia} critical value of stress intensity factor for crack arrest

K_{Ia}^* reduced crack arrest toughness $= K_{Ia}/\sqrt{10}$

K_{Ib} stress intensity factor due to bending stress

K_{Ic} critical value of stress intensity factor for crack initiation ('fracture toughness')

K_{Id} critical value of stress intensity factor for crack initiation under dynamic loading

K_{Im} stress intensity factor due to membrane stress

K_{IR} reference stress intensity factor (critical value)

ΔK_I range of applied stress intensity factor

K_{It} stress intensity factor due to thermal stresses

M_b correction factor due to bending stress

M_m correction factor due to membrane stress

M_t correction factor due to thermal stress

NDT   Nil Ductility Transition temperature

Q flaw shape parameter

R ratio of minimum/maximum stress in fatigue cycle

RT_{NDT} Reference NDT temperature

T_{CV} temperature for 50 ft lb and 0·035 in lateral expansion in Charpy V test

T_{NDT} nil ductility transition temperature determined from drop weight test

a depth of semi-elliptical flaw, or half depth of elliptical flaw

a_c minimum critical flaw size for normal conditions

a_f calculated end of life flaw size

a_i minimum critical flaw size for emergency and faulted conditions

c half length of flaw

da/dN rate of increase of flaw size per stress cycle (crack growth rate)

l length of flaw $(= 2c)$

n slope of $\log da/dN$ versus $\log \Delta K_I$ for crack growth

p crack penetration in crack arrest condition $(= 2a/t$ or $a/t)$

t section thickness

Note
Since ASME Codes use Imperial units and express temperatures in °F, this practice has been used in this chapter. Thus stress units are quoted as pounds per square inch or thousands of pounds per square inch (ksi).

1. INTRODUCTION

The *ASME Boiler and Pressure Vessel Code* Section III 'Nuclear Power Plant Components' (Division 1)[1]† is applicable to 'nuclear power plant items such as vessels, storage tanks, piping systems, pumps, valves and core support structures, and component supports, for use in, or containment of, portions of the nuclear power system of any power plant'. Further definition of a nuclear power system relates to 'the purpose of producing and controlling an output of thermal energy from nuclear fuel. ...' Although this definition is wide, in practice that part of Section III relevant to this chapter has been developed around the USA thermal reactor programme covering PWR (Pressurised Water Reactor) and BWR (Boiling Water Reactor) systems. Section XI 'Rules for In-service Inspection of Nuclear Power Plant Components' of the ASME Code consists of three Divisions, the relevant Division being 'Division 1—Rules for Inspection and Testing of Components of Light Water Cooled Plants',[2] thus reflecting the more specific application.

While the body of the Code is mandatory (if for those constructions where compliance with ASME rules is required, as indeed is called for by many licensing authorities) those parts of both Sections dealing with the assessment of flaws are contained in non-mandatory Appendices but are dependent upon the use of mandatory acceptance tests for materials and, in Section XI, can result in calculated critical sizes which are subject to mandatory consideration. In the latter case the mandatory statement is that 'The evaluation procedures and the acceptability criteria for these critical flaw parameters shall be the responsibility of the Owner and shall be subject to approval by the regulatory authority having jurisdiction at the plant site'—IWB-3600(C). However, again some licensing bodies regard all or part of these Appendices as mandatory. The flaw assessment method of Section III Appendix G is restricted to vessels and is effectively a demonstration of vessel integrity and a definition of operational conditions in the presence of postulated (rather than real) flaws. The method of Section XI Appendix A is in the main directed towards vessels but is defined as being applicable to ferritic materials of 4 in and greater thickness and of specified minimum yield strength not greater than 50 ksi; specifically it is stated that the concepts are not intended to apply to austenitic or high nickel alloys.

The concepts in both Sections have thus been developed essentially as means of assessment of flaws in thick walled ferritic steel pressure vessels for

† Unless stated otherwise references to the ASME Codes refer to the 1977 edition, published 1st July, 1977.

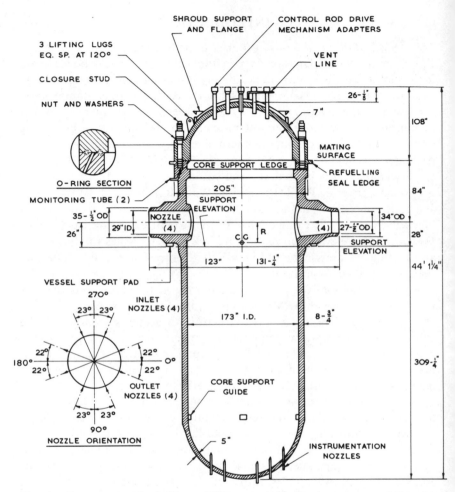

FIG. 1. Reactor vessel of 4-loop PWR system.[18] (Design pressure 2485 psig, operating pressure 2317 psig, design temperature 650 °F (343 °C), normal operating outlet temperature 618 °F (327 °C).

PWR and BWR systems. Since subsequent reference will be made to the more general validity of the concepts, some appreciation of the types of pressure vessel involved is pertinent. A typical design for a PWR and approximate dimensions, notably section thicknesses, is shown in Fig. 1. The wall thicknesses are generally of 6–8 in upwards and the steel adopted has been, almost invariably, a C–Mn–Ni–Mo steel (ASTM-A533B Class 1 and A508) with a specified minimum yield stress of $\sim 50\,000$ psi and ultimate tensile strength of 80 000 psi.

These large section thicknesses led to the development of linear elastic fracture mechanics (LEFM) for lower strength ferritic steels[3] and LEFM concepts have been adopted in both Sections of the ASME Code. Effectively, flaws in the vessels are regarded as behaving under plane strain conditions and their significance is assessed in accordance with the general relationship:

$$K_{Ic} = A\sigma\sqrt{\pi}\sqrt{a} \qquad (1)$$

where K_{Ic} is the fracture toughness under plane strain conditions or critical stress intensity factor,† A is a constant dependent upon flaw geometry and stress level, a is the significant dimension of the flaw, and σ is the nominal applied tensile stress perpendicular to the plane of the crack.

Detailed reviews of LEFM principles and methods have been made elsewhere (see, for example, references 4, 5 and 6). The unstable initiation and propagation of a flaw occurs when the summated stress intensity factors derived from all loadings attain the critical value K_{Ic}. This critical value is applicable for Mode I opening or plane strain conditions, its determination is precisely defined[7,8] and it is a temperature dependent material property. One important criterion in the definition of valid K_{Ic} determinations is that, to ensure plane strain conditions, the thickness B of the test specimen shall be related to the fracture toughness by $B \geq 2.5$ $(K_{Ic}/\sigma_y)^2$, where σ_y is the yield strength of the test piece at the temperature of the test. Thus, for the ferritic pressure vessel steel of 50 000 psi yield strength, the following minimum thicknesses of test piece will be required to give the respective valid K_{Ic} data:

B (in)	K_{Ic} (ksi$\sqrt{\text{in}}$)	B (in)	K_{Ic} (ksi$\sqrt{\text{in}}$)
1	31	6	77
2	45	8	90
4	63	12	109

In practice somewhat higher valid fracture toughness values may be obtained at a given thickness both because the sample tested may have higher than minimum yield strength and because of yield strength variation with temperature. A 12 in test piece of 70 ksi yield strength would permit a

---

† The stress intensity factor, K_{Ic} (or K_I for non-critical conditions), is not related to the 'stress intensity' used in stress calculations in Section III (Article NB-3000).

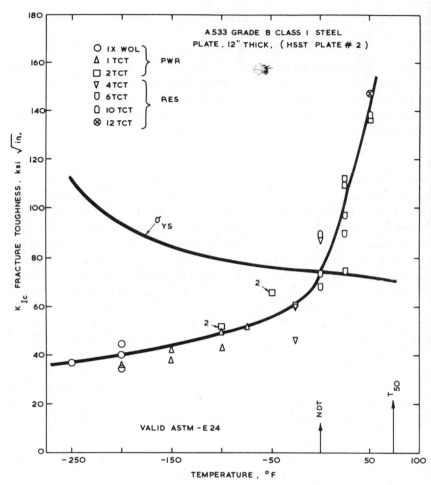

FIG. 2.   Temperature dependence of K_{Ic} for 12 in thick A533B Class 1 plate.[3]

valid toughness of 155 ksi $\sqrt{\text{in}}$ to be obtained. The variation of K_{Ic} with temperature for A533B steel is shown in Fig. 2,[3] in which the test piece thicknesses necessary to maintain plane strain conditions are indicated.

Acceptance of LEFM methods in the ASME Codes therefore implies either that plane strain conditions are maintained in all the section thicknesses considered or that an LEFM analysis will always give a conservative assessment of the critical size of defect.

2. DEVELOPMENT OF METHODS USED TO DERIVE FRACTURE TOUGHNESS IN ASME CODES

In common with most Design Codes for pressure vessels the earliest editions of ASME Section III Nuclear Power Plant Components based design against brittle or fast fracture on a transition temperature concept; an extension to give safe operating pressure/temperature conditions has been described by Porse[9] and was based upon the fracture analysis diagram due to Pellini.[10] None of these transition temperature concepts permitted any accurate quantification of the effect of flaws upon failure conditions although it is of interest to note that the current LEFM approach results, in spite of very different criteria, in recommended safe operating conditions which are similar to those of the transition temperature approach.

In 1971 a Task Group of the USA Pressure Vessel Research Committee (PVRC) of the Welding Research Council was charged with preparing recommendations for fracture toughness requirements for ferritic steels in nuclear power plants. The recommendations, published in Welding Research Council Bulletin No. 175,[11] formed the basis for revisions to Section III of the ASME Code which were first published in the Summer 1972 Addenda. Many of the methods, especially those dealing with derivation of fracture toughness data, have since been used in Section XI of the Code. Recommendations of the Task Group covered the reactor pressure vessel, reactor piping and bolting with Appendices covering material properties and methods of calculation of stress intensity factors. Where these have been adopted by the ASME Codes they are described later in this chapter. However, the method recommended for assessing the fracture toughness, K_{Ic}, of steels avoided direct measurement but involved correlations of material properties and merits particular consideration.

The size of test piece necessary to maintain plane strain conditions, to the upper shelf value of ~ 170 ksi$\sqrt{\text{in}}$ (Fig. 2), noted in the introduction to this chapter, is impracticable in terms of either quality control tests or even in terms of load capacity of most test machines. Thus to obtain an assessment of fracture toughness the Task Group recommended the use of Charpy V and drop weight test data and their correlation with a new concept of 'reference fracture toughness' K_{IR}. These reference values of fracture toughness were obtained by plotting dynamic initiation fracture toughness values K_{Id} and crack arrest fracture toughness values K_{Ia} against the test temperature relative to the nil ductility transition (NDT) temperature and defining a lower bound curve (Fig. 3). The data related exclusively to A533 Grade B Class 1 plate and A508 forgings. The lower bound curve (K_{IR}) was

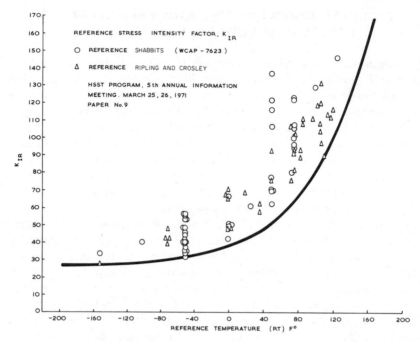

FIG. 3.   Derivation of curve of reference stress intensity factor (K_{IR}).[11]

effectively defined by the values of K_{Id} and K_{Ia}, the valid K_{Ic} values at a given temperature being higher (compare Fig. 2—where NDT temperatures generally are 0 °F–10 °F—with Fig. 3) and hence did not influence this lower bound. It is of interest to note that the dynamic initiation values were obtained using up to 8 in CKS test pieces and K values of up to 10^4/sec; since no criteria were specified for checking the validity of dynamic tests for initiation, those for static tests were adopted and the elevation of yield stress by dynamic loading permitted 'valid' toughness values of $\sim 200\,\mathrm{ksi}\sqrt{\mathrm{in}}$ to be obtained with this size of test piece. The K_{Ia} values were largely those produced in tests by Crosley and Ripling[12] in which normal initiation of a dynamic crack was produced in a contoured DCB test piece of 2 in thickness and the K_{Ia} value was calculated from the position at which the crack stopped in the test piece. Since that time (1971) considerable further effort has been expended on the development of a crack arrest criterion and the assessment of whether or not K_{Ia} is a material property. The question is especially relevant to the crack arrest criterion used in ASME XI Appendix A and is the subject of continuing investigations.[13]

The K_{IR} versus temperature relation to NDT curve was defined by:

$$K_{IR} = 26 \cdot 777 + 1 \cdot 223 \exp\{0 \cdot 0145[T - RT_{NDT} + 160]\} \qquad (2)$$

and is used in this form in the 1977 edition of Section III Appendix G of the ASME Code (Fig. 4). Although it is defined on dynamic initiation and crack arrest values, it is used in Section III (and in particular conditions in Section XI Appendix A) to define initiation toughness values and hence represents a

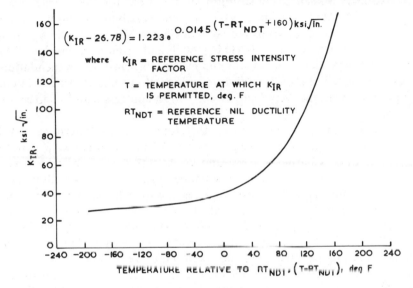

FIG. 4. K_{IR} versus $(T$-R$T_{NDT})$ relationship of Appendix G.

considerable degree of conservatism, compared with the use of directly measured K_{Ic} values. Reference to the NDT was shown as a conservative approach by comparison between dynamic yield strength σ_{yd} and K_{Id}. From the drop weight test the value K_{Id} at NDT temperature can be expressed as $K_{Id} = (\text{factor}) \times \sigma_{yd}$ where the factor varied between $0 \cdot 634$ and $0 \cdot 78$. The K_{IR} (K_{Id}) value of $39 \cdot 21 \, \text{ksi} \sqrt{\text{in}}$ recommended at NDT temperature corresponded to a factor of only $0 \cdot 55$ of dynamic yield stress.

To enter the K_{IR}/temperature relationship it is necessary to have determinations of NDT temperature from drop weight tests. Additionally, the PVRC Task Group recommended the use of Charpy V test data to define the reference temperature RT_{NDT}. From examination of test data it was concluded that a Charpy V notch specification based on lateral expansion of the test piece after fracture provided 'an automatic means for obtaining a constant level of fracture toughness irrespective of strength variations for material within a range of tensile properties in a specification or irrespective of deliberate changes in the specified tensile properties'. Lateral expansion (assumed to be equivalent to lateral contraction on which some investigations had been based, but measurable more readily) was selected in preference to energy absorbed requirements since the latter was strongly influenced by strength level and did not necessarily reflect changes in fracture toughness, especially the possibility of lower levels of toughness in the upper shelf. Indexing of Charpy V values to the K_{IR} curve was done at a temperature of NDT + 60 °F and, from relationships of K_{Id} with Charpy V lateral expansion and energy absorbed, recommendations were made for specified minimum values of 40 mils (0·040 in) lateral expansion or alternatively 35 mils (0·035 in) and not less than 50 ft lb energy absorbed.

The recommended procedure, therefore, was to determine NDT temperature, T_{NDT}, by drop weight tests; at a temperature of T_{NDT} + 60 °F conduct three Charpy V tests and, if lateral expansion values were ≥ 40 mils, then T_{NDT} is the reference temperature R T_{NDT}. If 40 mils was not achieved in any one of the three test pieces then further tests were to be made to determine the temperature T_{40} at which all three values were ≥ 40 mils: R T_{NDT} was then defined as $T_{40} - 60$ °F. The alternative of 35 mils and 50 ft lb energy absorbed was also regarded as 'reasonable'. Both drop weight and Charpy tests were to be taken at quarter thickness ($t/4$) location and to cover base metal, heat affected zone and weld metal. It was further recommended that the required properties must be met throughout the life of the component, particularly taking into account neutron irradiation damage. Although the precise values in the Charpy V recommendations have since been modified to some extent, the overall recommendations have been adopted in acceptance values in NB-2300 'Fracture Toughness Requirements for Material' of the (mandatory) Section III of the ASME Code.

Overall the recommendations of the PVRC Task Group provided the basis for yielding a conservative estimate of LEFM fracture toughness, avoiding the difficulties of direct measurement and depending upon the commonly used Charpy V and drop weight acceptance tests. The principles are used virtually unchanged in the demonstration of integrity in the presence of postulated flaws detailed in Section III Appendix G but have been modified for some operational conditions in Section XI Appendix A.

3.   THE METHODS OF ASME III APPENDIX G

The Appendix G approach is essentially that of applying LEFM to postulated surface defects of simple shape in a pressure vessel; these defects are generally of depth equal to one quarter of the material thickness and of length equal to one and a half times the material thickness. The summation of the stress intensity factors resulting from the various loadings is required to be less than the K_{IR} value of the material (in several instances a safety factor being included). Appendix G gives methods for defining the material toughness, the calculation of stress intensity factors and defines the limits of applicability of the method.

3.1.   Fracture Toughness

The fracture toughness K_{IR} of materials having a specified minimum yield strength at room temperature of 50 ksi or less is obtained from the K_{IR} versus $(T - RT_{NDT})$ plot of Fig. 4 ('unless higher K_{IR} values can be justified'). The derivation of RT_{NDT} is given in ASME Section III NB-2300. Charpy V tests are taken at $t/4$ depth with an orientation (for plate material) in a direction normal to the principal rolling direction, other than the thickness direction, and with the notch normal to the surface of the material. Drop weight tests may have any orientation but orientation and location have to be reported in the Certified Material Test Report. Similar requirements are made for weld metal tests.

The reference temperature RT_{NDT} is established by:

(1)   Determining a temperature at or above the nil ductility transition temperature T_{NDT} by drop weight tests.

(2)   At a temperature not greater than $(T_{NDT} + 60°F)$ Charpy V test pieces shall give at least 35 mils (0·035 in) lateral expansion and not less than 50 ft lb absorbed energy. Retesting is permitted and, if these values are achieved, then T_{NDT} is the reference temperature RT_{NDT}.

(3)   If the latter requirements are not met then additional Charpy V tests are made to determine the temperature T_{CV} at which these energy and lateral expansion measurements are met. The reference temperature is then defined as $T_{CV} - 60°F$. An alternative of deriving T_{CV} from a full impact curve is also permitted.

These tests are applied to the base material of the primary acceptance tests, the base material, the heat affected zone and weld metal from weld procedure tests and the weld metal deposits used to define weld metal properties.

For materials of specified minimum yield stress greater than 50 ksi but less than 90 ksi the relationship of Fig. 4 may be used provided that fracture mechanics data are obtained on at least three heats of the material including weld metal and heat affected zone. The data required relate to dynamic tests similar to the K_{Id} data used in the original assessment by the PVRC Task Group and they must show values equal to or above those shown in Fig. 4. Additionally, where neutron radiation may influence properties, the effect of radiation on the K_{IR} curve has to be determined for the material prior to its use in manufacture.

3.2. Defect Size

For section thicknesses of 4 in to 12 in the surface defect size postulated is $t/4$ depth (corresponding to the position of the material acceptance tests) $\times$ $1\frac{1}{2}t$ length. For sections greater than 12 in thickness the size is limited to 3 in depth $\times$ 18 in and for less than 4 in thick sections the size is taken as 1 in depth $\times$ 6 in. The use of smaller sizes of postulated defects is permitted 'if a smaller size of maximum postulated defect can be assured'. The postulated defects are all assumed to be normal to the direction of maximum stress.

3.3. Calculation of Stress Intensity Factors

Definition of the defect as a semi-elliptical surface defect of fixed aspect ratio ($a/2c = \frac{1}{6}$) permits the use of relatively simple expressions for the calculation of stress intensity factors. The stress intensity factor is given by $K_{\mathrm{I}} = M \times$ membrane stress where M is a factor shown graphically in Fig. 5 and incorporating the effects of the ratio of the applied stress relative to yield stress of the shape of the defect (constant) and of the material thickness. (A full description of the derivation of these values is given in Appendix 3 of the PVRC Task Group Report.[11]) Thus: for membrane tension the stress intensity factor is given by $K_{\mathrm{Im}} = M_{\mathrm{m}} \times$ membrane stress; for bending stress $K_{\mathrm{Ib}} = M_{\mathrm{b}} \times$ maximum bending stress where M_{m} and M_{b} are taken from Fig. 5. Similarly the stress intensity factor K_{It} for a radial thermal gradient is given by $K_{\mathrm{It}} = M_{\mathrm{t}} \times$ F, where F is the temperature difference through the wall (in °F), and M_{t} has the values shown in Fig. 6. (Note that this relates to a simplified analysis for an infinitely long defect and is applicable only to temperature gradients of a particular form (Fig. G-2214-3 of Appendix G) and when the temperature change starts from a steady state condition and has a rate less than about

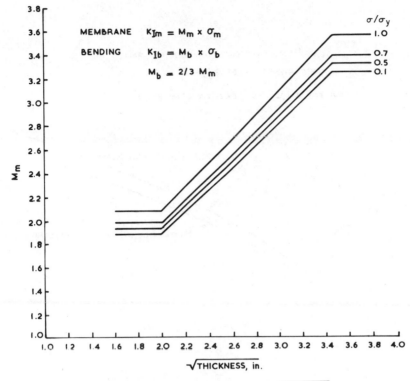

FIG. 5. Variation of M_m, M_b with $\sqrt{\text{thickness}}$.

100°F per hour; for other conditions alternative methods must be used for calculating the effective K_{It}.)

3.4. Allowable Operational Conditions

Reactor operating conditions are usually defined as normal, upset, emergency or faulted conditions, the stress applied generally increasing with progression through these conditions. Early editions of Appendix G referred to these conditions. The 1977 edition refers to 'Service Limits' at Level A, B, C or D, precise definitions of which are given in NA-2140 of ASME Section III. Generally they correspond to the earlier nomenclature, Level D Service Limits, for example, 'permit gross general deformations with some consequent loss of dimensional stability and damage requiring repair, which may require removal of the component from service'. The recommendations of ASME III Appendix G relate only to Level A and

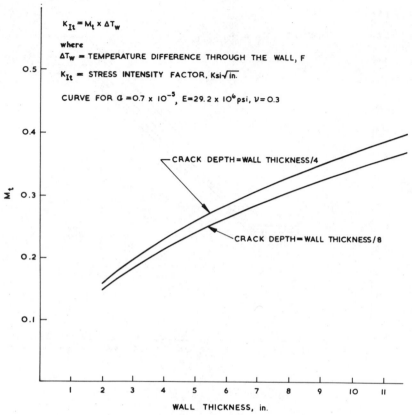

$$K_{It} = M_t \times \Delta T_w$$

where

ΔT_w = TEMPERATURE DIFFERENCE THROUGH THE WALL, F

K_{It} = STRESS INTENSITY FACTOR, Ksi√in.

CURVE FOR $G = 0.7 \times 10^{-5}$, $E = 29.2 \times 10^6$ psi, $\nu = 0.3$

CRACK DEPTH = WALL THICKNESS/4

CRACK DEPTH = WALL THICKNESS/8

M_t

WALL THICKNESS, in.

FIG. 6.   Variation of M_t with thickness.

Level B Service Limits, the range of conditions of Level C and Level D Service Limits being regarded as 'too diverse to allow the application of definitive rules . . .' although the use of similar principles is recommended. The procedure for Level A and Level B Service Limits distinguishes between areas of nominal stress and areas in or near stress concentrations.

The nominal stress areas, defined as 'Shells and Heads Remote from Discontinuities', are subject to the restriction that, for any condition of temperature and pressure, the requirement $2K_{Im} + K_{It} < K_{IR}$ is satisfied. This introduces a safety factor of 2 to the stress intensity factor produced by primary stresses. When the elastically calculated stresses exceed the yield strength of the material the above methods of calculation of stress intensity factors are invalid. Reference is made to the procedure recommended by the

PVRC Task Group for calculating defect significance under these conditions. The method is limited to areas of local yielding only (i.e. enclosed by elastic material) and entails modification of the elastically calculated stress to a value approximating to the real stress by taking into account the effect of yield in the material.

The areas of more complex stress, defined as 'Nozzles, Flanges, and Shell Regions near Geometric Discontinuities' require the introduction of bending stresses. Again all stress intensity factors due to primary stresses (including those caused by bolt pre-loading) are increased to $2 \times$ actual value, giving a requirement of:

$$\underbrace{2K_{\text{Im}} + 2K_{\text{Ib}}}_{\text{due to: primary stresses}} + \underbrace{K_{\text{Im}} + K_{\text{Ib}}}_{\text{secondary stresses}} < K_{\text{IR}}$$

Thermal stresses are regarded as secondary but the method of derivation for K_{It} used in Fig. 6 is not recommended for these areas.

Nozzles are excluded from this type of assessment on the grounds that a quantitative evaluation 'is not feasible at this time', although it is noted that the critical size of defect in the nozzle region may be significantly less than that in the vessel shell. Reference is also made to an approximate method, due to Gilman and Rashid,[14] for defects on the inside corner of a nozzle; the method was described by the PVRC Task Group.[11]

For hydro-test conditions the requirements to be met are:

$$\underbrace{1 \cdot 5K_{\text{Im}} + 1 \cdot 5K_{\text{Ib}}}_{\text{due to: primary stresses}} + \underbrace{K_{\text{Im}} + K_{\text{Ib}}}_{\text{secondary stresses}} < K_{\text{IR}}$$

The safety factor for this condition is thus reduced to $1 \cdot 5 \times$ compared with the $2 \times$ for Levels A and B Service Limits. The hydro-test is recommended to be made at a temperature not less than $RT_{\text{NDT}} + 60 \,^\circ F$, the vessel thus having a minimum K_{IR} of $57 \, \text{ksi} \sqrt{\text{in}}$ compared with $\sim 170 \, \text{ksi} \sqrt{\text{in}}$ at normal ($\sim 600\,^\circ$F) operating temperatures. Since the applied pressure at hydro-test may be $1 \cdot 25 \times$ design temperature, the hydro-test effectively gives a greater ratio of applied stress intensity factor to K_{IR} than operation, the severity depending upon the actual material toughness levels at the test temperature.

3.5. Summary

The ASME III Appendix G approach to the assessment of defects in pressure vessels is based on LEFM methods for a postulated or 'design' defect of $\frac{1}{4}t \times 6t$. The defect size is assumed to be constant throughout vessel life. Safety factors are used in the assessment and further conservatism is introduced by the use of K_{IR} values rather than the K_{Ic} values more pertinent for initiation assessments. Material toughness values are generally derived from correlations with Charpy V notch and drop weight test data, rather than by direct measurement and, from the rigid requirements of ASME III Article NB-2300, encompass plate or forging, weld metal and heat affected zone. References are made to behaviour throughout the life of the component, including the effect of neutron irradiation, but no guidance is given in Appendix G on how to assess or estimate changes in material properties which could occur during reactor life.

The assessment is based on Level A and Level B Service Limits and does not detail methods for Level C and Level D Service Limits which are likely to give a more severe test of integrity. Specific recommendations are not made for the regions subject to the more complex stresses. The material fracture toughness properties at an operating temperature of about 600 °F are not given but there is the implicit assumption that the toughness remains at values not less than the upper value of ~ 170 ksi$\sqrt{}$in given in Fig. 4. Overall, the Appendix should be viewed as a procedure for obtaining allowable pressure vessel loadings for normal service conditions even in the presence of a substantial defect. The probability of defects of such a size existing has to be assessed against the Section III requirement for volumetric inspection of weld areas by radiography with a sensitivity of 1–2% of thickness and the ASME XI requirements for in-service inspection.

4.   THE METHODS OF ASME XI APPENDIX A

The methods recommended in ASME XI Appendix A to assess the significance of defects are similar in principle to those of ASME III Appendix G. For example, LEFM methods are used and fracture toughness values are generally derived by correlation with other tests. However, since the methods are concerned with real rather than postulated defects, more elaborate treatments are necessary.

ASME XI requires the periodic inspection of parts of the reactor vessel by ultrasonic examination. All indications that exceed the standards of ASME XI IWB-3500 are considered as planar defects (including porosity and slag) and Appendix A gives recommended methods for analysis of such indications. Since the indications can represent flaws of varying shape and orientation a more conventional fracture mechanics approach has to be applied than was possible with the restricted defect size and shape considered in ASME III Appendix G. Thus: the defect is resolved into a shape capable of analysis; relevant stresses for all conditions in the reactor life are determined; appropriate stress intensity factors are calculated; material fracture toughness values are derived; LEFM methods are used to calculate crack growth and critical sizes of defects; the acceptability of the defect is compared with the criteria of Section XI IWB-3600.

4.1. Characterisation of Defect

The procedure translates the indications from ultrasonic examination into simplified shapes, amenable to LEFM analysis, taking account of proximity of flaw to surface of material and to other flaws and of flaw orientation. The rules for these procedures are in the body of the Section XI Code, IWA-3300, and are, therefore, mandatory.

The indications from ultrasonic examination are circumscribed by a rectangular or square area the lengths of which are taken as the axes of ellipses or semi-ellipses covering the area of the indication. Although this inscribed ellipse need not cover all of the indication, the method facilitates the application of developed LEFM methods and gives a conservative assessment of flaw size. The flaw is described by length l and depth a for a surface flaw or $2a$ for a buried flaw (Figs. 7 and 8).

Individual characterised flaws are considered to be in a plane normal to the component surface with the major axis parallel to the component surface, thus effectively giving the loading conditions used in LEFM analyses. Fracture mechanics principles have been applied to further re-characterisation of flaws, for example to allow for the proximity of flaws to a free surface or to another flaw. The stress intensity magnification factor was used to define the need for re-characterisation; for subsurface flaws the criterion of 1·05 stress intensification factor resulted in treatment as a surface flaw if the nearest approach to the surface was less than one half of their depth. Thus in Fig. 7 Defect No. 4 when $s \leq d$ the depth a is given by $(2d + s)$ rather than $2d$; a similar approach for multiple flaws is illustrated in Fig. 9.

Similar basic principles are applied when indications are found on inclined planes (the flaws being projected into planes, normal to maximum principal stresses—in some cases necessitating assessment for severity on two planes) and for parallel planar flaws. Overall the characterisation procedure provides a conservative assessment of flaw size and orientation

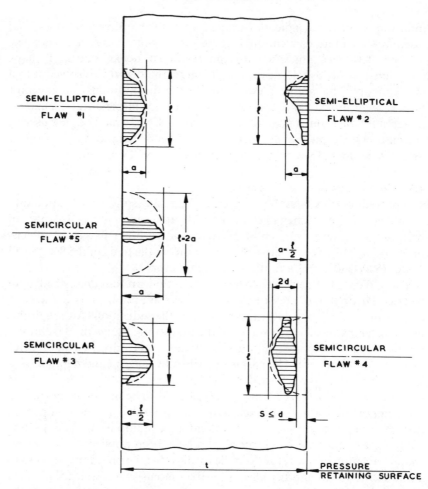

ILLUSTRATIVE FLAW CONFIGURATIONS AND DETERMINATION OF DIMENSIONS a AND l

FIG. 7. Surface planar flaws oriented in plane normal to pressure retaining surface.

to give a flaw of effective size of $l \times a$ (or $2a$) when used in LEFM procedures.

4.2. Calculation of Stress Intensity Factors

Since assessment must be made for a variety of shapes of flaws, a more general derivation is necessary than that used in ASME III Appendix G to determine stress intensity factors. Although more sophisticated methods

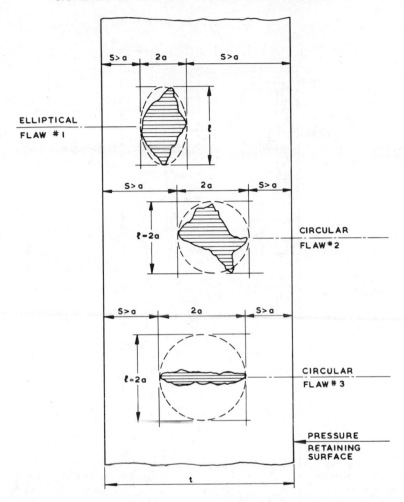

ILLUSTRATIVE FLAW CONFIGURATIONS AND DETERMINATION OF DIMENSIONS 2a AND ℓ

FIG. 8.   Sub-surface planar flaws oriented on plane normal to pressure retaining surface.

are permitted (and indeed it is acknowledged that the recommended equation may be inadequate for complex geometries and stress distributions), the recommended equation is:

$$K_I = \sigma_m M_m \sqrt{\pi} \sqrt{a}/Q + \sigma_b M_b \sqrt{\pi} \sqrt{a}/Q \tag{3}$$

where σ_m and σ_b are membrane and bending stresses; a is the minor half

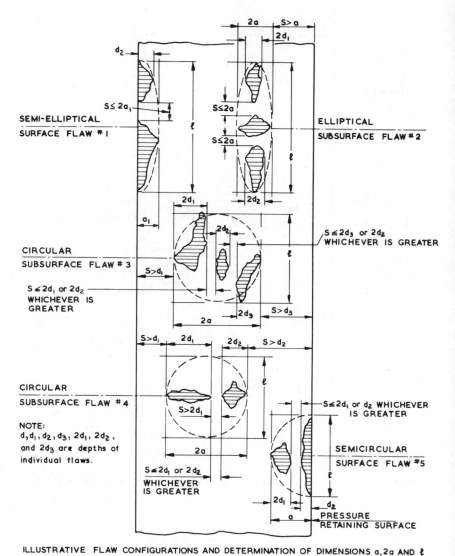

ILLUSTRATIVE FLAW CONFIGURATIONS AND DETERMINATION OF DIMENSIONS a, 2a AND ℓ

FIG. 9. Multiple planar flaws oriented in plane normal to pressure retaining surface.

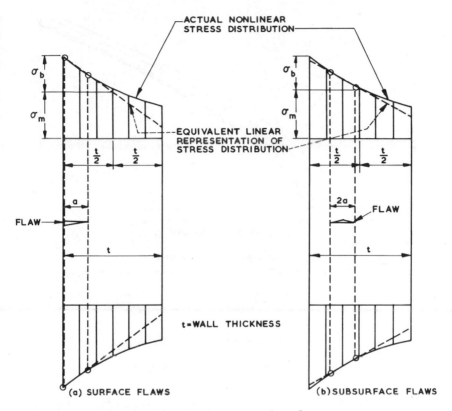

Fig. 10.   Linearised representation of stresses.

diameter of embedded flaw (flaw depth for surface flaw); Q is the flaw shape parameter as determined from Fig. A-3300-1 (Fig. 11) using $(\sigma_m + \sigma_b)/\sigma_y$ and the flaw geometry; M_m is a correction factor for membrane stress (Fig. A-3300-2 (Fig. 12) for subsurface flaws, Fig. A-3300-3 for surface flaws); and M_b is a correction factor for bending stress (Fig. A-3300-4 for subsurface flaws, Fig. A-3300-5 (Fig. 13) for surface flaws). (Note that M_m and M_b differ numerically and in function from those used in Appendix G.)

The stresses at the flaw location are resolved into membrane and bending stresses across the wall thickness and all forms of loading are considered, namely pressure stresses, thermal stresses, discontinuity stresses and residual stresses. For non-linear stress distributions a technique is given whereby the stress can be represented linearly over the depth of the defect and broken down into membrane (σ_m) and bending stresses (σ_b) (Fig. 10).

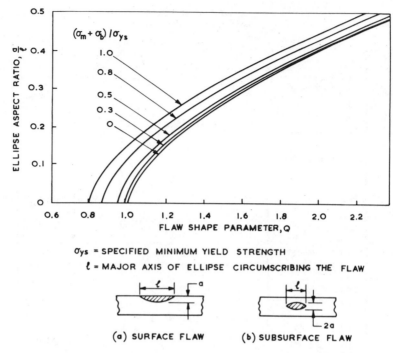

FIG. 11.   Shape factors for flaw model.

The use of the flaw shape parameter Q (Fig. 11) and membrane (Fig. 12) or bending (Fig. 13) stress correction factors represent normal LEFM procedures developed for specific geometries (see, for example, references 15, 16 and 17).

4.3. Assessment of Material Properties

The recommendations for assessing fracture toughness differ from the Appendix G approach in two respects. Firstly, distinction is made between K_{Ic} and K_{Ia} and, secondly, preferred values are those from the specific material with the lower bound values, correlated with RT_{NDT}, representing an alternative approach. The prime requirement is that K_{Ic} and K_{Ia} should represent lower bound, or conservative, values preferably from the specific material and product form involved. It is further recommended that 'the values so used should be justified on the basis of current technology, and should take into account material variability, testing techniques, and any other variables which may lower these toughness values'.

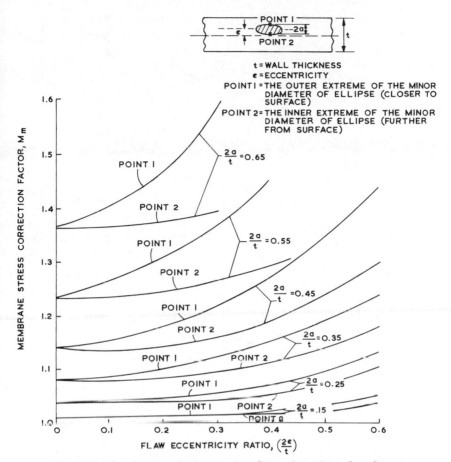

FIG. 12.   Membrane stress correction factor for sub-surface flaws.

The alternative lower bound correlation with RT_{NDT} (determined according to the rules of NB-2331) distinguishes between K_{Ia} and K_{Ic}. The recommended values (Fig. 14) are 'intended to be very conservative' since direct measurement of fracture toughness is the recommended procedure. The K_{Ia} versus $(T - RT_{NDT})$ relationship is identical to that of K_{IR} versus $(T - RT_{NDT})$ used in Appendix G. The K_{Ic} relationship is given by:

$$K_{Ic} = 33 \cdot 2 + 2 \cdot 81 \exp \{0 \cdot 02(T - RT_{NDT} + 100)\} \qquad (4)$$

The upper fracture toughness values for both K_{Ic} and K_{Ia} are shown as $\sim 200 \, \text{ksi} \sqrt{\text{in}}$ compared with that for K_{IR} shown in Appendix G as

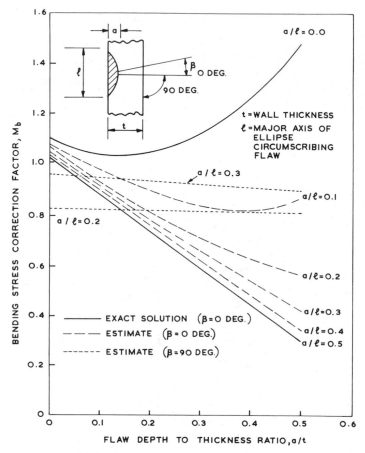

FIG. 13. Bending stress correction factor for surface flaws.

~ 170 ksi$\sqrt{\text{in}}$; comments are made on the relevance of these values in the Discussion to this chapter.

Specific recommendations are made for assessing the effect of neutron irradiation on K_{Ic} and K_{Ia}. As with the evaluation of irradiated toughness values the primary recommendation is for measurement from surveillance test pieces, but when such data are not available provision is made for estimating changes in RT_{NDT} from trend curves (Fig. 15). These curves are again 'intended to be very conservative' and especially they highlight the deleterious effect of Cu on susceptibility to irradiation damage. Where Cu

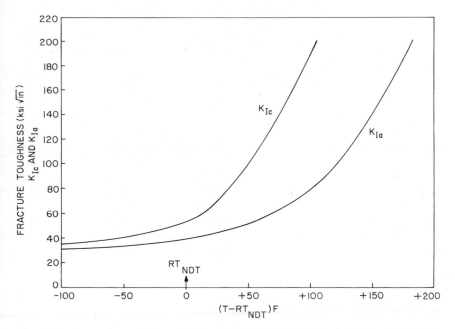

FIG. 14.   Lower bound K_{Ic} and K_{Ia} test data for SA-533 Grade B Class 1, SA-508
Class 2 and SA-508 Class 3 steels.

content is not known the recommendation is to assume the highest copper
shown.

To estimate the result of fatigue crack growth on a specific flaw the
recommendation is for the use of the Paris law:

$$\frac{da}{dN} = C_0(\Delta K_I)^n \qquad (5)$$

where da/dN is the rate of change of crack depth per cycle; C_0 is a constant;
n is the slope of $\log(da/dN)$ versus $\log \Delta K_I$ plot; and ΔK_I is the range of the
applied stress intensity factor.

Again the recommendation is for direct data determination from the
actual material and product form taking into account 'material variability,
environment, and other variables that affect the data'. When such data are
not available upper bound curves are given (Fig. 16) distinguishing between
surface flaws (which may be subject to a water environment) and subsurface
flaws. The effect of environment has been shown to have a marked effect
upon crack growth rates at a given ΔK value. Growth rate is also influenced

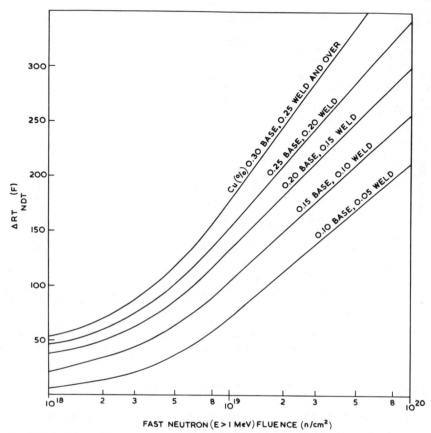

FIG. 15.   Effect of fast neutron fluence and copper content on shift of RT_{NDT} for
reactor vessel irradiation at 550 °F.

by frequency of test, R (the ratio of minimum to maximum K in the cycle),
the form of the cycle, etc. Some limitation due to these effects has been
made in the Summer 1977 Addendum to Appendix A by restricting
application of the relationship for surface flaws to R values between 0 and
0·25. A full description of these effects is given by Mogford in Chapter 7 and
indicates that these curves of Appendix A (Fig. 16) may not always be
conservative in conditions of a water environment.

4.4. Analysis of Flaws
Different forms of analysis are required for normal (including upset) and
for emergency and faulted conditions. The normal (and upset) conditions

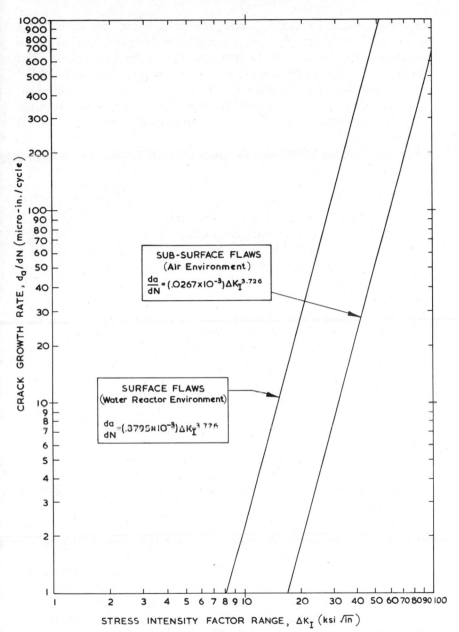

Fig. 16. Upper bound fatigue crack growth data for SA-533 Grade B Class 1, SA-508 Class 2 and SA-508 Class 3.

are those transients occurring during the testing and operation of the reactor and which are taken into account in design. A typical list of transients which are anticipated during the life of a PWR pressure vessel is given in Table 1.[18] The emergency and faulted conditions are those defined as low probability incidents which may prevent subsequent operation of the plant and the only essential requirement is that of safe shutdown of the plant. These definitions may be compared with those of Level A, B, C and D Service Limits of ASME III NA-2140.

For normal conditions the procedure recommended is that of a cumulative crack growth calculation and determination of the critical size. The change of applied stress intensity (ΔK_I) for a given transient is used to calculate the amount of growth produced on the characterised flaw using the procedures outlined above. The incremental growth Δa so calculated is added to the depth of the characterised flaw a. This procedure is repeated for each of the remaining design transients of the vessel, taking them in approximate chronological order, so producing an end of life flaw size (a_f).

The critical size for all normal conditions is calculated taking into account the effect of neutron irradiation. The stress intensity factors are calculated for flaws geometrically similar to that of the observed flaw but with values of a bounding the actual value of a_f. Thus, for a given transient, the critical size may be calculated by comparison of the applied stress intensity with the material toughness (taking into account temperature and irradiation effects). The material toughness value used is K_{Ia}, equivalent to K_{IR}. By calculation of the critical size for all transients the minimum value found then represents the critical flaw size (a_c) for normal (and upset) conditions. (The form of equation precludes simpler procedures to establish critical sizes.)

For emergency and faulted conditions initiation (K_{Ic}) and crack arrest (K_{Ia}) data are used. The end of life irradiation dose is determined as a variation through the thickness of the component at the position of the flaw and, from irradiated fracture toughness data, the K_{Ic} and K_{Ia} profiles are then established through the wall thickness as a function of time after start of the transient. Stress intensity factors are calculated for various depths of flaw bounding the calculated end of life depth. The extent of flaw penetration at which the calculated stress intensity factor exceeds the K_{Ic} profile through the thickness corresponds to the critical size for initiation (a_i) and the size for arrest (a_a) is given by intersection with the K_{Ia} curve (Fig. 17). The procedure is repeated for each transient and for a variety of times after the start of each transient. The minimum value of a_i obtained by this procedure, and for which the crack arrest penetration (p) is greater than

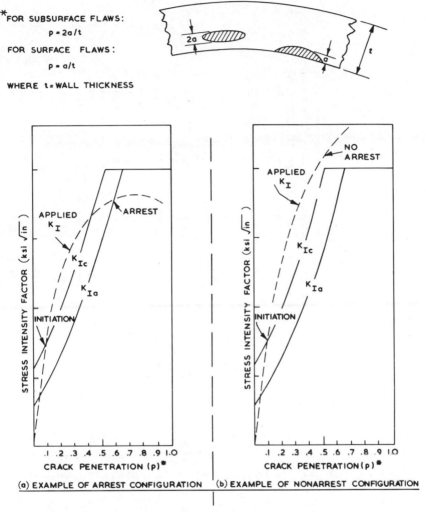

FIG. 17. Determination of critical flaw sizes for postulated conditions.

0·75 (regarded as 'non-arrest') is taken as the critical size for emergency and faulted conditions. (Emergency and faulted conditions may result in cold water being injected on the inner surface of the vessel; this will reduce the surface temperature and hence the fracture toughness and produce a high tensile thermal stress. All of these factors will vary with time and the result will be to produce the type of stress intensity factor profiles shown in Fig. 17.)

TABLE 1

REACTOR VESSEL DESIGN TRANSIENT SUMMARY[18]

Design transient	Class	Lifetime occurrences	Duration Δt (min)	Pressure change, ΔP (psi)	Temperature change, ΔT (°F)	(Cold leg) minimum (°F)
1. Plant heat up at 100°F/h	Normal	200	300	2 332	487	70
2. Plant cool down at 100°F/h	Normal	200	300	2 332	487	70
3. Unit loading/unloading 0–15% full power	Normal	500	35	Nil	16	549
4. Plant loading at 5% full power/min	Normal	13 200	22	66	10	547
5. Plant unloading at 5% full power/min	Normal	13 200	27	66	5	556
6. Step load increase of 10% full power	Normal	2 000	6	40	5	553
7. Step load decrease of 10% full power	Normal	2 000	7	100	7	556
8. Large step decrease with steam dump	Normal	200	20	245	14	550
9a. Steady state fluctuations A	Normal	150 000	2	50	6	554
9b. Steady state fluctuations B	Normal	3×10^6	6	12	1	556.6
10. Feedwater cycling at hot shutdown	Normal	2 000	90	64	24	533
11. Loop out of service normal shutdown	Normal	80	10	156	18	542
12. Loop out of service normal startup	Normal	70	10	44	5	554
13. Refuelling	Normal	80	10	Nil	108	32
14. Boron concentration equalisation	Normal	26 400	60	25	Nil	
15. Loss of load without immediate turbine or reactor trip	Upset	80	>2	866	42	557
16. Loss of power	Upset	40	>150	595	30	555
17. Partial loss of flow	Upset	80	>3.5	385	13	549

No.	Description	Condition					
18.	Reactor trip	Upset					
	(a) No cool down		230	>2	280	7	555
	(b) Cool down, No SI		160	>1	650	40	519
	(c) Cool down and SI		10	>10	750	100	457
19.	Inadvertent depressurisation	Upset	20	>17	1813	110	457
20.	Inadvertent startup—inactive loop	Upset	10	>3	357	21	538
21.	Control rod drop	Upset	80	>3	420	25	535
22.	Inadvertent safety injection	Upset	60	>17	342	17	544
23.	Small loss of coolant accident	Emergency	5	400	1080	525	
24.	Small steam break	Emergency	5	67	390	260	
25.	Complete loss of flow	Emergency	5	4	450	41	
26.	Feedwater line break	Faulted	1				
27.	Reactor coolant pipe break	Faulted	1				
28.	Large steam line break	Faulted	1				
29.	Reactor coolant pump, locked rotor	Faulted	1				
30.	Control rod ejection	Faulted	1				
31.	Turbine roll test	Test	20	>17	625	110	447
32.	Primary side hydrostatic test before startup at 3 107 psig and at $RT_{NDT} + 60°F$	Test	10		3 107	a	70
33a.	Primary side leak tests at 2 485 psig at 250°F minimum	Test	200		2 500	a	70
33b.	Secondary side leak tests at 2 485 psig at 250°F minimum	Test	80		30	a	70

a Temperature change ΔT negligible.

4.5. Acceptability of Flaws

The requirements of ASME XI include in-service inspection and pre-service inspection using techniques to be applied during the in-service inspection. Initially the pressure vessel will have been inspected to ASME III standards using radiography as the means of volumetric inspection. The acceptance standards for defects in ASME III are, as in most pressure vessel design codes, based on what can be achieved with good practice; thus they reflect quality assurance requirements rather than fracture mechanics requirements related to the service conditions and integrity of the vessel. Acceptability standards in ASME XI (Article IWB-3500) have the same intention and it is only when ultrasonic indications exceed these standards that 'analytical procedures, such as that described in (ASME XI) Appendix A', are required to assess critical flaw sizes.

The Acceptance Standards IWB-3500 in ASME cover allowable planar indications for both surface and subsurface flaws and laminar flaws. The values were selected on the basis of $\frac{1}{10}$ of the 'design flaw' used in ASME III Appendix G for Level A and B Service Limits; as the stress intensity factor varies according to the square root of flaw depth, the safety factor may be regarded as 3, this having been claimed as being consistent with the principal design factor (UTS/3) of Section III.

The evaluation of flaws exceeding these levels (by the methods of Appendix A or an approved alternative method) results in critical values (defined in ASME XI Article IWB-3600) as:

a_f the maximum size to which the detected flaw is calculated to grow during the remaining service lifetime of the component;

a_c the minimum critical flaw size of the indication under normal operating conditions;

a_i the minimum critical flaw size of the indication for initiation of non-arresting growth under postulated design, emergency and faulted conditions.

These parameters 'shall not exceed the following acceptability criteria:

$$a_f < 0 \cdot 1 a_c, \qquad a_f < 0 \cdot 5 a_i$$

or, alternatively:

$$a_f < a_c^*, \qquad a_f < 0 \cdot 5 a_i$$

where a_c^* is calculated using the reduced fracture toughness parameter K_{Ia}^*, where $K_{Ia}^* = K_{Ia}/\sqrt{10}$'.

The reduced fracture toughness factor effectively introduces a safety factor of $\sqrt{10}$ on toughness, rather than of 10 on defect size; it avoids some of the anomalies which may be introduced with specific variations of stress intensity factor variations through a section. This option and the use of specific safety factors will be discussed later in more detail.

4.6. Summary

In contrast to ASME III Appendix G, ASME XI Appendix A uses an LEFM approach for defects found in a structure. The methods of resolving 'indications' (usually found by ultrasonic examination) into ellipses or semi-ellipses for which fracture mechanics methods are available are mandatory and are similar to those adopted in other methods of analysis.[19,20] The non-mandatory Appendix gives methods for the estimation of stress intensity factors but acknowledges that in some cases these methods may be inadequate and permits the use of more sophisticated methods.

The use of directly measured fracture toughness values is preferred but, as in Appendix G, correlations with drop weight and Charpy V test data are acceptable. Distinction is made between values for initiation (K_{Ic}) and for crack arrest (K_{Ia}) (equivalent to K_{IR}), the former being used only in the assessment of fault and emergency conditions. Direct measurement of the effect of irradiation upon properties and of crack growth data, including the effect of environment, is recommended but alternative methods are given to estimate properties.

Crack growth due to changes in operational stresses is calculated giving a value of defect size a_f; its acceptability is then checked, distinguishing between normal and upset conditions and fault and emergency conditions. For normal and upset conditions the calculated size at end of design life a_f is compared with the critical value a_c; this is obtained using K_{IR} (K_{Ia}) values, corrected for irradiation effects, and is based on a geometrically similar flaw and evaluation of all the stress intensity factors for normal and upset transients to determine the most severe. A similar procedure for determining critical values is applicable to fault and emergency conditions but distinction is made between crack initiation (K_{Ic}) and crack arrest (K_{Ia}) values. For each transient these toughness values are plotted as a variation through the thickness of the component at a series of times following the start of the transient. By superimposition of the stress intensity factor on to these plots, values for crack initiation (a_i) and crack arrest (a_a) can be derived. The critical size for initiation is that for a non-arresting crack or for one in which crack arrest would not be predicted before the crack had penetrated 75% of the section thickness.

The mandatory section of ASME XI specifies that the calculated flaw sizes at end of life (a_f) are acceptable if, for normal conditions a_f is less than $\frac{1}{10}$ of a_c and, for fault and emergency conditions, a_f is less than $\frac{1}{2}$ of a_i.

5. DISCUSSION

The methods used to assess the significance of defects are those of linear elastic fracture mechanics, with respect both to materials properties and definition of stress intensity factors. For thick sections and low temperature stressing after irradiation, several situations will exist where defect extension will occur under elastic conditions. For thinner sections and/or for operation in the upper shelf regime (above $RT_{NDT} + 200\,°F$) varying degrees of yielding may occur before crack extension and hence the methods depend upon LEFM being a conservative approach under such conditions (i.e. LEFM will predict a critical size of defect equal to or smaller than that necessary to cause failure). For many situations this conservatism will exist but checks to ensure that this is the case are advisable; one method of checking using a plastic collapse criterion is described by Darlaston in Chapter 4.

The initiation toughness K_{Ic} measured by ASTM E-399 methods[7] is about 150–160 ksi$\sqrt{\text{in}}$ for a 12 in thick test piece.[3] The highest value measured for dynamic or arrest toughness is ~ 200 ksi$\sqrt{\text{in}}$[11] and the fracture toughness ($K_{Ic} + K_{Ia}$) curves of Appendix A are terminated at this value; the K_{IR} curve of Appendix G is terminated at 170 ksi$\sqrt{\text{in}}$. The implication is made that toughness remains constant at these values up to pressure vessel operating temperatures. Under these upper shelf conditions LEFM is not applicable and more rigorous analyses would require the use of elastic–plastic fracture mechanics.[21] As will be seen from later chapters, at present elastic–plastic concepts are not yet at a stage of universal acceptance and require either a re-definition of design criteria, using J or COD for example, or, more commonly, using a value of K_{Ic} calculated from test data on small test pieces not meeting the ASTM validity criteria for LEFM. Equivalent energy and J methods are used most often. In this upper shelf ductile regime, crack extension (initiation) first occurs by tearing (or slow crack growth) and instability in a test piece may not occur until the applied load, displacement or energy is considerably greater than that at which first crack extension occurred. Thus fracture toughness criteria based on criteria taken at maximum load in a test piece may yield apparent K_{Ic} values of 400 ksi$\sqrt{\text{in}}$[22] compared with initiation values of less than

$200\,\text{ksi}\sqrt{\text{in}}$ on a similar type of material.[23] Progress is being made to determine the significance of tearing or slow crack growth[24] but distinguishing a critical criterion over such a wide range of apparent toughness requires further development before confidence can be placed in any value other than that for crack initiation, i.e. the onset of slow crack growth. The development of elastic–plastic fracture mechanics will permit a better definition of upper shelf behaviour and a better definition of critical defect size under these conditions. A similar conclusion was made by the PVRC Task Group in 1972.

The assumption of constant toughness under upper shelf conditions is debatable. Some test data indicate a reduction in toughness between the onset of upper shelf toughness and pressure vessel operating temperature. Tests, using initiation criteria, show reductions from $190–160\,\text{ksi}\sqrt{\text{in}}$[23] (Fig. 18), similar work attributing the reduction to dynamic strain ageing.[25] This type of behaviour highlights the need for development of definitive methods for fracture toughness determinations at all temperatures encountered in service.

The ASME XI Appendix A recommends that 'conservative' values of K_{Ic} and K_{Ia} can be obtained from the specific materials used and the values 'should be justified on the basis of current technology'. The equipment necessary for such determination of K_{Ic} in thick section is not available in many test houses and the additional material requirements to establish fully the material properties would be extensive; hence the great majority of applications will be dependent upon the lower bound values derived from correlations with RT_{NDT}.

A large amount of conservatism is introduced, particularly in the 'transitional' temperature range by the use of K_{IR} and K_{Ia} relationships to assess conditions for fracture initiation. However, the distinction in ASME XI Appendix A between K_{Ic} and K_{Ia} used in assessing the significance of defects under fault and emergency conditions has less substantiation. The crack arrest toughness values are based upon a particular test piece form and, when this approach was introduced, there had been relatively little work to verify that K_{Ia} was a material property or to verify the concept of a crack arrest situation under the stressing conditions found under faulted or emergency conditions. More recently there has been a significant increase in research and development of crack arrest criteria as a material property[13] and investigations of structural behaviour;[26] the overall evidence for validity remains less than that for fracture initiation criteria. The conditions for which this crack arrest concept (giving a penetration of less than 75 % the section thickness) would be applicable may be relatively few in number.

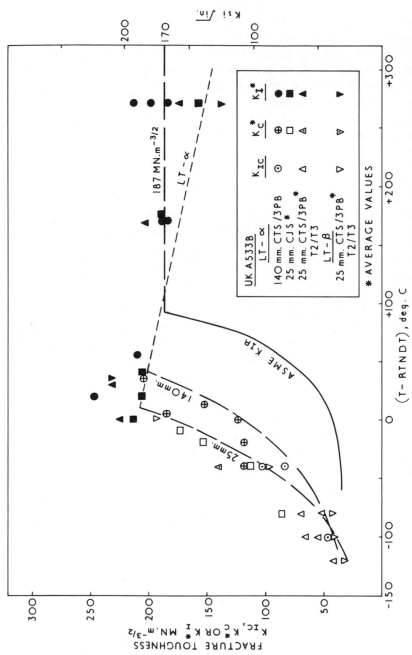

Fig. 18. Comparison of fracture toughness data for A533 Grade B Class 1 steel of UK origin with the ASME K_{IR} curve and a constant upper shelf plateau of 170 ksi $\sqrt{\text{in}}$.[23]

In discussing several examples of application of Appendix A, Cooper[27] comments that 'with the possible exception of a few cases involving abnormal fluence effects, our impression is that this particular evaluation condition will not be significant except for extremely large flaws'.

The categorisation of flaws (Appendix A) results in shapes amenable to analysis by well-defined LEFM methods taking into account, for example, the elevation of stress intensity by adjacent flaws to yield a conservative (over-large) definition of flaw size. Perhaps the greater uncertainties in this area are the assumption of constant aspect ratio $(a/2c)$ when considering flaw growth and the actual crack growth rates recommended. Experimental work[28,29] has shown that the aspect ratio may vary during fatigue crack growth, the variation being dependent upon the relative magnitude of bending and tension stresses. The resultant variation in stress intensity factor is likely to result in appreciable discrepancies only when $a/2c$ is small. The Appendix A recommendation for direct measurement of crack growth rate, including the effect of environment, is even less likely to be attempted than that for direct measurement of fracture toughness. The uncertainties in the use of the rates recommended in Fig. 16 are discussed in Chapter 7.

The use of Appendix G results in a demonstration of integrity by defining operational conditions applicable when hypothetical defects of $\frac{1}{4}t$ depth are present in specific areas. Appendix A considers real defects found by non-destructive examination and, generally, calculates their significance on a conservative basis. The significance of the uncertainties noted above has to be seen against the mandatory safety factors required by ASME XI on the calculated critical sizes. For normal operating conditions the calculated end of life flaw size a_f must be less than $0 \cdot 1 a_c$ or less than a_c^* (defined previously). The value of $0 \cdot 1 a_c$ introduces an appreciable safety factor, presumably intended to take into account uncertainties of stress analysis, definition of stress intensity factor, crack growth rates, material toughness, etc. The safety factor was based on a factor of 3 on toughness throughout life but, as noted by Cooper[27] it can result in anomalies in analysis. When thermal stresses are present the stress intensity factor may be larger for a small flaw than for a large flaw and the stress intensity factor of the small flaw may exceed one third of the material toughness even though the flaw size is less than one tenth of the critical size. The use of the reduced fracture toughness value $K_{1c}/\sqrt{10}$ overcomes this and other anomalies associated with the $0 \cdot 1 a_c$ criterion. Overall, the safety factor incorporated for normal operational conditions appears to give adequate compensation for uncertainties in material knowledge or stressing conditions. For faulted and emergency conditions the end of life size a_f must be less than $0 \cdot 5 a_i$. This reduced safety factor for conditions beyond which continued operation is not required is dependent upon crack arrest criteria; further development and proving of this concept is necessary before the significance of any safety factor can be appreciated fully.

Application of the methods of Appendix A is complex and liable to several pitfalls. The greatest difficulties may arise because of lack of relevant data but variations in stress intensity factor solutions for a given geometry and the complexity of crack growth are also likely to cause problems. The difficulties have been recognised and examples of calculations[30] are commended to anyone contemplating a similar exercise.

The more general application of Appendix A values and methods to other structures, such as bridges, vessels not fabricated in accordance with ASME III, pipework, etc., is valid but introduces a number of difficulties. In any application of fracture mechanics to an existing structure uncertainties are immediately raised because little or no knowledge exists of material toughness and stress analyses are often inadequate. LEFM toughness values of ferritic steels are, with few exceptions, limited to A533B and allied steels. The ASME correlations are based on data from these steels (and represent the only substantial amount of any form of fracture toughness data which has been published for low strength steels) and are quoted as being relevant to steels of specified minimum yield strength of 50 ksi or less. The steels tested often had actual yield strength values of 60 ksi or greater and data on other steels of similar strength level confirm that Charpy V/K_{Ic} correlations could be valid over a wider range.[20] ASME III Appendix G indicates that similar correlations could be applicable for steels up to 90 ksi yield strength but does not offer evidence of such correlations. Thus, for many steels in common use, a reasonable conservative estimate of fracture toughness may be obtained which will permit a first estimate to be made of defect significance.

In contrast to the requirements for many fabrications, ASME III requires a detailed stress analysis to be made (and limits stress levels, geometries, etc. which are permissible). This stress analysis forms an essential element in fracture mechanics assessments which is often not easily available for assessments of other structures. The need for accurate stress analysis is paramount in any fracture mechanics assessment. However, Cooper[27] has noted the inadequacy of some ASME III stress reports which have the objective of demonstrating that Code requirements are satisfied; this may be achieved by conservative analyses which result in the inability to evaluate defect significance until more refined analysis is done. For other structures the problem is magnified several-fold.

In general the LEFM methods of Appendix A may be applied to other structures to give a conservative definition of critical defect size. The problems noted above should be considered and especially (since most structures are of thinner section and LEFM conditions will be less relevant) General Yielding Fracture Mechanics methods[20] and plastic collapse[19] must be taken into account. With these provisos the LEFM methods offer a useful method of assessment of significance of defects in fabricated structures, especially when viewed against the usual background of non-acceptability of crack-like defects.

In conclusion it is emphasised that the relevant Appendices of the ASME Codes are non-mandatory. Effectively they recommend the use of LEFM methods for assessing the significance of defects, postulated or real, and aim at providing conservative assessments. With the possible exception of the use of a crack arrest concept, they achieve this aim. For real defects (ASME XI Appendix A) the use of directly determined material properties and any adequately substantiated fracture mechanics method is advocated; alternative methods for derivation of material properties by correlation or from conservative assessments and for analysis are given. The only mandatory requirements are for material properties (in ASME III, giving data amenable to correlation), for defect sizes determined by in-service inspection which must be subject to LEFM analyses, for characterisation of such defects and for permissible end of life defect sizes relative to critical sizes (ASME XI). Even though they are non-mandatory the Appendices represent a significant advance in being the first in any Code which advocate methods permitting the assessment of significance of defects in ferritic steels.

ACKNOWLEDGEMENTS

Acknowledgement is made to the American Society of Mechanical Engineers for permission to reproduce figures taken from *ASME Boiler and Pressure Vessel Codes* and to make quotations from the text of the Codes.

Figure 18 is reproduced by permission of the Council of the Institution of Mechanical Engineers from 'Conference on Tolerance of Flaws in Pressurised Components'.

REFERENCES

1. *ASME Boiler and Pressure Vessel Code*, Section III, 'Nuclear Power Plant Components', Division 1. 1977 Edition. American Society of Mechanical Engineers, New York.

2. *ASME Boiler and Pressure Vessel Code*, Section XI, 'Rules for In-service Inspection of Nuclear Power Plant Components', Division 1. 1977 Edition. American Society of Mechanical Engineers, New York.
3. WESSEL, E. T. LEFM for thick walled welded pressure vessels: material property considerations. In: *Proc. Symposium on Fracture Toughness Concepts for Weldable Structural Steels*, 1969, Chapman and Hall, London.
4. 'Symposium on fracture toughness testing and its applications'. *ASTM STP* 381, 1965.
5. 'Plane strain crack toughness testing of high strength metallic materials'. *ASTM STP* 410, 1966.
6. KNOTT, J. F. *Fundamentals of Fracture Mechanics*, 1973, Butterworth, London.
7. *ASTM Standards*. Standard method of test for plane strain fracture toughness of metallic materials. E 399-72, 1972.
8. *BS* 5447: 1977, 'Methods for plane strain fracture toughness (K_{Ic}) testing', British Standards Inst., London.
9. PORSE, L. *Trans. ASME, J. Bas. Eng.*, 1964, **86**, 743.
10. PELLINI, W. S. and PUZAK, P. P. WRC Bull., No. 88, 1963, Welding Res. Council, New York.
11. WRC Bull. No. 175. PVRC recommendations on toughness requirements for ferritic materials, 1972, Welding Res. Council, New York.
12. CROSLEY, P. B. and RIPLING, E. J. Crack arrest fracture toughness of A533B Class 1 pressure vessel steel. HSSTP-TR-8, 1970, Oak Ridge National Lab.
13. 'Fast fracture and crack arrest'. *ASTM STP* 627, 1977.
14. GILMAN, J. D. and RASHID, Y. R. 'Three dimensional analysis of reactor pressure vessel nozzle'. G/2 Conf. on Structural Materials in Reactor Technology, 1971, Berlin.
15. TADA, H., PARIS, P. and IRWIN, G. *The Stress Analysis of Cracks Handbook*, 1973, Del Research Corporation, Pa.
16. SIH, G. C. *Handbook of Stress Intensity Factors*, Leigh Univ., 1973, Bethlehem, Pa.
17. ROOKE, D. P. and Cartwright, D. J. *Compendium of Stress Intensity Factors*, 1976, HMSO, London.
18. *UKAEA. An Assessment of the Integrity of PWR Pressure Vessels*, 1976, HMSO, London.
19. DARLASTON, B. J. L. A CEGB two-criteria proposal. This volume, Chapter 4.
20. BURDEKIN, F. M. The British Standard Committee WEE37 Draft and IIW approaches. This volume, Chapter 3.
21. PARIS, P. C. Fracture mechanics in the elasto-plastic regime, flaw growth and fracture. *ASTM STP* 631, 1977, **3**, 27.
22. WITT, F. J. and MAGER, J. R. A procedure for determining bounding values on fracture toughness K_{Ic} at any temperature. ORNL-TM-3894, 1972.
23. INGHAM, T. and SUMPTER, J. D. G. Design against fast fracture in thick walled pressure vessels. *Proc. Conf. Tolerance of Flaws in Pressurised Components*. Held in May 1978. Inst. Mech. Engineers, London.
24. PARIS, P. C., TADA, H., ZAHOOR, A. and ERNST, H. The theory of instability of the tearing mode of elastic–plastic crack growth. NUREG-0311, 1977. Material Research Lab., Washington Univ., St Louis.

25. OSTENSSON, B. In: *IAEA Symp. on Application of Reliability Techniques to Nuclear Power Plants 'Fracture toughness of A533B steel at elevated temperatures'*. To be published, 1978, IAEA-SM-218/17. IAEA, Vienna.
26. CHEVERTON, R. D. Pressure vessel fracture studies pertaining to a PWR LOCA-EGC thermal shock: experiments TSE-1 and TSE-2. ORNL/NUREG/TM-31, 1976.
27. COOPER, W. E. and MOY, G. 'ASME XI flaw evaluation procedures and applications to nozzles'. Teach-in ASME XI, 1976. UKAEA in collaboration with ASME and Inst. Mech. Engineers.
28. IIDA, K. and KAWAHARA, M. Univ. Tokyo, Naut. Report 9011, 1975.
29. HODULAK, L., KORDISCH, H., KUNZELMANN, S. and SOMER, E., Institute für Festkörpermechanik, Freiburg, 1976.
30. MARSTON, T. U. (Ed.). 'Background and Application of ASME Section XI Appendix A' prepared by ASME Flaw Evaluation Working Group EPRI Report No. NP-719-SR. Pub. Electric Power Research Institute, Palo Alto, USA, 1978.

Chapter 3

THE BRITISH STANDARD COMMITTEE WEE/37 DRAFT AND IIW APPROACHES

F. M. BURDEKIN

University of Manchester Institute of Science and Technology, UK

SUMMARY

For some years there have been discussions in the UK as to whether there should be a formal British Standard on the acceptability of flaws in welded construction. There is still no clear agreement as to whether rules for the acceptance of flaws in welded construction should be given in the form of a Standard or Code and at the present time it seems likely that information on this subject will be issued in the form of a guidance document rather than a mandatory Standard

At the same time discussions have been held within the Commissions of the International Institute of Welding and a guidance document prepared by a Working Group of Commission X dealing with assessment of the significance of weld defects in respect of brittle fracture failure.

This chapter sets out the background and detailed requirements to the British Standard WEE/37 discussions and the IIW approach giving the basic information necessary for assessment of the acceptability of flaws as detailed in these approaches.

1. BACKGROUND

Many laboratories and research establishments in the United Kingdom, Europe, the United States, Japan and elsewhere have investigated the effects of flaws on components subject to stress. This work has been carried out by both theoretical and experimental investigations and has looked at all types of engineering and structural components. An attempt has been made to take into account as much of this work as possible in two parallel documents setting out methods for the derivation of acceptance of weld flaws. These documents are the British Standard WEE/37 Committee document[1] and the International Institute of Welding Commission X document.[2]

Amongst the leading sources of work on the effect of weld defects on structural performance has been the work carried out for many years at the Welding Institute (formerly the British Welding Research Association). In the early period much of this work concentrated on the effects of defects under fatigue loading, since it was found that weld defects often had a much more deleterious effect under fatigue loading than under static loading. It was possible to show from this work the relative severity of different types of weld defect in steel, aluminium and other materials under fatigue conditions.

At the same time research programmes have been carried out over many years at the Welding Institute into the problem of fracture in welded structures. This work led naturally to a study of the general subject of fracture mechanics applied to welded fabrications, structures, pressure vessels, etc. The research programmes were influenced strongly by examination of service failures, and by general developments in the study of fracture mechanics. The influence of weld defects on the general fracture behaviour of joints became part of these studies.

From this work came the realisation that many existing codes and standards, which set acceptance limits for defects in welds, were based arbitrarily on factors other than whether the defects would affect the structural integrity and service performance of the component. It was considered that this situation merited open debate in the relevant industries, between customers and plant operators, contractors, fabricators, inspection authorities and research workers. Pressures were becoming apparent from fabricators that they were being asked to repair the most minor of defects, and as methods and standards of non-destructive testing improved, smaller and smaller defects could be detected. The situation had arisen that no codes were available which enabled the contracting parties involved in purchasing and manufacturing welded components to decide on the acceptability of weld defects on a fitness for purpose basis.

In 1967[3] and 1968,[4] two international conferences were held in London on the significance of weld defects. Amongst the papers, there was included one paper at each conference which summarised the research work carried out at the Welding Institute.[5,6] The second of these papers, by Harrison, Burdekin and Young[6] was one of the first attempts to set down methods for determining the significance of weld defects on a fitness for purpose basis.

In 1969 the British Standards Institution set up a committee (WEE/-/4) to report back whether it was considered that there was now sufficient information available to enable a British Standard to be prepared on acceptance levels for weld defects on a fitness for purpose basis. This committee reported that such a Standard could be prepared, and in due course a further British Standard committee was set up to carry out the task of preparing such a document (the WEE/37 committee). From about 1970 to 1975 this committee deliberated on the precise wording and technical details of the Standard until eventually a 'Draft Standard' was issued for industrial comment in February 1976.[1] This document followed largely a development of the philosophy of the Welding Institute work. As a result of the comments received it was decided that modifications were necessary to the Draft, and that it would then be published as a guidance document by the British Standards Institution rather than as a mandatory Standard.

Over the same period from 1970 onwards, Commission X of the International Institute of Welding set up a Working Group to prepare a document on acceptance levels for weld defects with respect to brittle fracture failure. This Working Group had representatives from the United Kingdom, Japan, Sweden, Hungary, Belgium, France, West Germany, and its recommendations were submitted to the full IIW Annual Assembly and published in *Welding in the World* in 1975.[2]

From this background it should be noted that the WEE/37 approach is concerned with acceptance levels for weld defects in most forms of welded construction, and considers the effect of weld defects on the full range of expected modes of failure. The scope of the approach is restricted to defects in fusion welded joints in ferritic steels, austenitic steels and aluminium alloys of 10 mm thickness and above, although the principle of the approach may be applied to other materials and thicknesses when data are available. The IIW document is concerned only with failure by brittle fracture, but uses a similar basic approach to the WEE/37 methods. Since some further developments have taken place in the WEE/37 approach since the time of publication of the IIW document, the details described below will be the latest position in each case.

Both of the approaches deal with the whole field of welded construction, and are not restricted to the field of pressurised components. The approaches do consider, however, those factors relevant to pressurised components, so that they are fully relevant to the question of flaw analysis for pressurised components. The field of elastic–plastic fracture mechanics is one of continuing development, and modifications to the details of the methods may well take place in some areas. Nevertheless the approaches are considered to present a satisfactory method of determining the acceptability of weld flaws, on a fitness for purpose basis, without requiring complex and expensive analyses. The approaches apply in principle to defects other than weld flaws, although originally designed and presented in terms of weld flaws.

It is recommended that those without experience in this general field should seek guidance and assistance where necessary to confirm that any assumptions made and conclusions drawn are valid for the particular circumstances.

2. THE BASIC APPROACH

When assessing the effect of weld defects on a fitness for purpose basis, it is not possible to provide a single list of defects which will cause failure. Whether or not a particular defect matters, will depend upon its type, size, location and orientation, the materials in which it is situated, and the stress field and environment to which it is subjected.

The implication of defect assessment on a fitness for purpose basis is thorough examination after any post-weld heat treatment and/or proof test by non-destructive testing using techniques capable of locating and sizing defects in critical areas. The limitations of non-destructive testing methods have to be taken into account.

It is normal practice for there to be two levels for defect assessment. The quality control level is set based on what can be achieved by normal good practice and workmanship, and the fitness for purpose level on whether the defects will prevent the component fulfilling its design requirements throughout its design life. Thus the fitness for purpose assessment methods have three basic functions:

(1) To assess the acceptability of a particular defect found during construction or service, based on whether this will affect service performance.

(2) To enable guidance to be given to particular applications Standards to enable general acceptance levels for defects to be set for different types of component.

(3) To enable acceptance levels to be set for specific components in a

particular contract in advance of manufacture and service, which are more realistically linked to the effect of defects on performance than are normal quality control levels.

2.1. Types of Defect

The WEE/37 and IIW documents cover the effects of the following weld defects:

planar defects —cracks
        —lack of fusion or penetration
        —undercut, root undercut, root concavity, etc.
non-planar defects—cavities, pores, etc.
        —solid inclusions

The method of assessment of the group of defects termed planar defects is by fracture mechanics analyses. The method of assessment for non-planar defects is based upon examination of a wide range of experimental data, supported by fracture mechanics concepts where relevant.

2.2. Modes of Failure

Whilst the IIW document is concerned, under its terms of reference, only with failure by brittle fracture, the WEE/37 document considers the effects of the following modes of failure:

(1) Brittle fracture.
(2) Fatigue.
(3) Yielding due to overloading of remaining cross-section.
(4) Leakage in containment vessels.
(5) Corrosion, erosion, corrosion fatigue, stress corrosion.
(6) Instability (buckling).
(7) Creep/creep fatigue.

It was considered by the WEE/37 committee that the information available to assess the effect of defects on failure by brittle fracture and fatigue was sufficiently well understood that detailed methods of treatment could be given. For the other modes of failure, the effect of defects is sometimes less, and in some cases complex effects of the environment are also involved. These other modes of failure are treated by general guidelines in so far as the effect of weld defects is concerned.

2.3. Information Required for Assessment

In order to be able to derive the effects of different types of defect on all of the different modes of failure, relevant data in the following areas will be required:

(1)   Position, size, nature, and orientation of defect.
(2)   Structural and weld geometry.
(3)   Stresses (however arising) and temperatures including transients.
(4)   Conventional yield or proof stress in tension, and modulus of elasticity.
(5)   Fatigue/corrosion fatigue S/N and crack propagation data.
(6)   Fracture toughness (K_{Ic} or δ_c) data.
(7)   Creep rupture, creep crack propagation and creep fatigue data.
(8)   Bulk corrosion and stress corrosion cracking (K_{ISCC}) data.

If the method is being used to assess the acceptability of a known defect in a particular location, the final assessment will probably be either a comparison of actual with allowable defect sizes, or a comparison of stress intensity factor at the defect tip with fracture toughness. If the method is being used to determine allowable defect sizes in a general situation it will be necessary to consider the various alternative sizes/shapes of defect which are equivalent in effect. This is discussed further below.

2.4. Stresses

The notation and definition of stresses adopted in the WEE/37 and IIW documents is that used in the *ASME Boiler and Pressure Vessel Code*, Section III.[7] These stresses may be summarised as follows:

(1)   Average primary stress, P_m.
(2)   Primary bending stress, P_b.
(3)   Secondary stress, Q (including thermal and residual stress).
(4)   Peak stress, F.

In these documents, the peak stress, F, is defined as the *increment* of stress which is additive to the primary plus secondary stress due to local discontinuities. Peak stresses do not cause distortion but have to be considered as possible sources for the initiation of failure. Where stresses in peak regions are defined by an elastic stress concentration factor, K_t, the peak stress, F, is defined as $(K_t - 1)P_m$.

The stresses to be considered are those in the neighbourhood of the defect, but should not take into account the stress concentration effect at the tip of the defect caused by the defect itself as this is covered in the assessment.

For planar defects the component of stress normal to the plane of the defect should be used. To apply this the defect should be resolved onto a plane normal to the most significant principal stress.

3. DETAILED ASSESSMENT FOR BRITTLE FRACTURE

The assessments for brittle fracture failure are firstly divided into three sections, planar defects, non-planar defects, and defects arising from imperfect shape.

The assessments for planar defects are based upon fracture mechanics methods. There is, however, a further sub-division for planar defects, between calculating the acceptability of known defects, and estimation of tolerable sizes of planar defects in a general situation. For known defects the method of treatment for stress levels below yield is based upon linear elastic fracture mechanics[8-10] in the WEE/37 document, and the method for stress levels above yield is based upon the crack opening displacement (COD) concepts of post-yield fracture mechanics.[11-13] The estimation of tolerable sizes of planar defects is treated by the COD post-yield fracture mechanics. The use of other forms of elastic–plastic fracture mechanics, such as J-contour integral, two-criteria, or equivalent energy concepts amongst others in this rapidly developing field, is admitted as an alternative if agreed by all parties involved, but again expert advice is recommended in such circumstances.

3.1. Defect Dimensions and Interaction

In these assessments, planar defects are idealised by the height (t) and length (l) of the rectangle which contains them (Fig. 1). The actual fracture mechanics analyses used for part thickness defects are in fact based upon the elliptical family of defects which would be contained within a rectangle of those dimensions (i.e. minor axis t, major axis l).

Guidance is given on the question of interaction of neighbouring defects. The criterion adopted is that where the presence of a second defect would cause an elevation of 20 % in the stress intensity factor at the tip of the first defect, an equivalent defect is assumed to be effective having height and length of the rectangle containing both defects. The resultant criteria and revised dimensions for interacting defects are shown in Figs. 2(a) and (b).

For surface and buried defects, it is important to check that the ligaments at the surfaces are stable. This problem arises mainly with long part thickness defects if net section yielding or local plastic collapse in pressure vessels could occur. In such cases rules are given to indicate whether buried

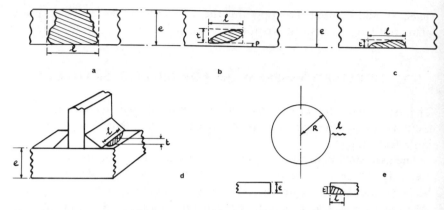

FIG. 1.   Defect dimensions. (a) Through-thickness defect, required dimensions e, l.
(b) Embedded defect, required dimensions e, l, t, p. (c) Surface defect, required
dimensions e, l, t. (d) Defect at toe of weld, required dimensions e, l, t. (e) Defect at
hole, required dimensions e, l, R.

flaws should be re-categorised as surface flaws and whether surface flaws
should be re-categorised as through thickness flaws.

3.2. Known Defects—Below Yield Stress

The basic method is to calculate the stress intensity factor (K_I) from the
defect dimensions and stress levels and compare this to the fracture
toughness (K_{Ic}). Provided the stress intensity factor is less than $0.7 K_{Ic}$, the
defect is considered acceptable. In most cases this implies a safety factor of 2
on defect size, but this can reduce to 1.25 for long part thickness flaws.

For part thickness defects the stress intensity factor is calculated from
standard published linear elastic fracture mechanics solutions[14–20] and the
formula:

$$K_I = \frac{\sqrt{a}}{Q_0}(\sigma_m M_m + \sigma_b M_b)$$

where $a = t$ for surface defects; $a = t/2$ for embedded defects; σ_m = tensile
component of stress; and σ_b = bending component of stress.

The values of σ_m and σ_b should be derived from the total stress
$P_m + P_b + Q + F$ in the area in which a defect is contained or where it is
required to determine an allowable flaw size.

The parameter Q_0 is the standard flaw shape factor, values of which are
shown in Fig. 3. The parameters M_m, M_b, are magnification factors for free

SCHEMATIC DEFECTS	CRITERION FOR INTERACTION	EFFECTIVE DIMENSIONS AFTER INTERACTION
1 — COPLANAR SURFACE DEFECTS 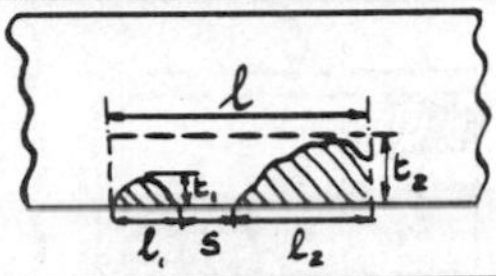	$s \leq \dfrac{l_1 + l_2}{2}$	$t = t_2$ $l = l_1 + l_2 + s$
2 — COPLANAR EMBEDDED DEFECTS 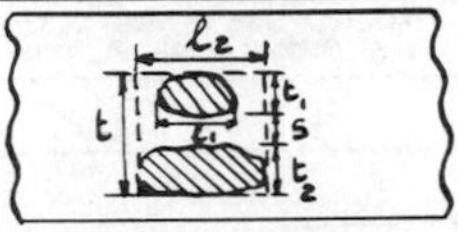	$s \leq \dfrac{t_1 + t_2}{2}$	$t = t_1 + t_2 + s$ $l = l_2$
3 — COPLANAR EMBEDDED DEFECTS 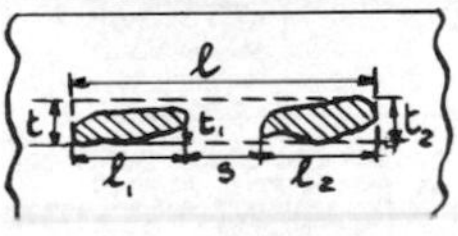	$s \leq \dfrac{l_1 + l_2}{2}$	$t = t_2$ $l = l_1 + l_2 + s$
4 — COPLANAR SURFACE AND EMBEDDED DEFECTS. 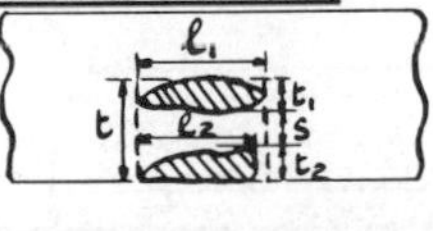	$s \leq \dfrac{t_1 + t_2}{2}$	$t = t_1 + t_2 + s$ $l = l_1$
5 — COPLANAR EMBEDDED DEFECTS 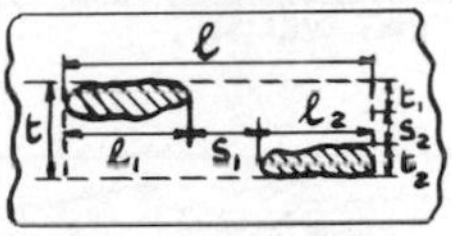	$s_1 \leq \dfrac{l_1 + l_2}{2}$ AND $s_2 \leq \dfrac{t_1 + t_2}{2}$	$t = t_1 + t_2 + s_2$ $l = l_1 + l_2 + s_1$
6 — SURFACE AND EMBEDDED DEFECTS 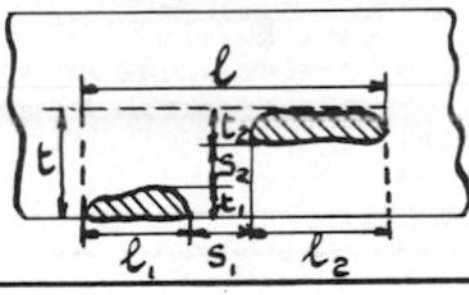	$s_1 \leq \dfrac{l_1 + l_2}{2}$ AND $s_2 \leq \dfrac{t_1 + t_2}{2}$	$t = t_1 + t_2 + s_2$ $l = l_1 + l_2 + s_1$

Fig. 2(a).   Planar defect interactions (coplanar).

SCHEMATIC DEFECTS

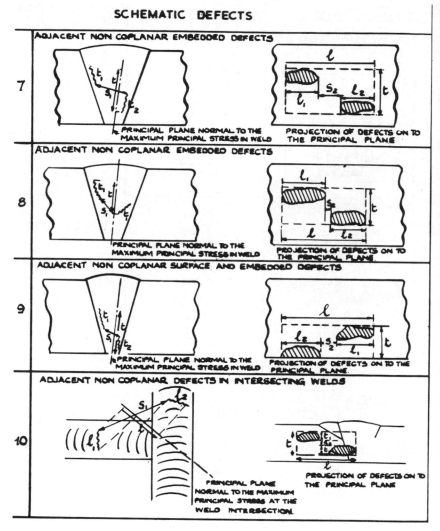

FIG. 2(a)—*contd.*

CRITERION OF INTERACTION	EFFECTIVE DIMENSIONS FOR INTERACTION
$S_1 \leq \dfrac{t_1 + t_2}{2}$ AND $S_2 \leq \dfrac{l_1 + l_2}{2}$ [l_1 & l_2 ARE PROJECTED LENGTHS]	t THE MINIMUM HEIGHT OF CONTAINMENT RECTANGLE CONSTRUCTED ON THE PLANE NORMAL TO THE MAXIMUM PRINCIPAL STRESS IN THE WELD $l = l_1 + l_2 + S_2$
$S_1 \leq \dfrac{t_1 + t_2}{2}$ AND $S_2 \leq \dfrac{l_1 + l_2}{2}$ [l_1 & l_2 ARE PROJECTED LENGTHS]	t THE MINIMUM HEIGHT OF CONTAINMENT RECTANGLE CONSTRUCTED ON THE PLANE NORMAL TO THE MAXIMUM PRINCIPAL STRESS IN THE WELD. $l = l_1 + l_2 + S_2$
$S_1 \leq \dfrac{t_1 + t_2}{2}$ AND $S_2 \leq \dfrac{l_1 + l_2}{2}$ [l_1 & l_2 ARE PROJECTED LENGTHS]	t THE MINIMUM HEIGHT OF CONTAINMENT RECTANGLE CONSTRUCTED ON THE PLANE NORMAL TO THE MAXIMUM PRINCIPAL STRESS IN THE WELD $l = l_1 + l_2 + S$
$S_1 \leq \dfrac{l_1 + l_2}{2}$ AND $S_2 \leq \dfrac{t_1 + t_2}{2}$ [t_1 & t_2 ARE PROJECTED HEIGHTS]	$t = t_1 + t_2 + s$ THE MINIMUM LENGTH l OF CONTAINMENT RECTANGLE IS CONSTRUCTED ON THE PLANE NORMAL TO THE MAXIMUM PRINCIPAL STRESS IN THE WELD.

FIG. 2(b). Planar defect interactions (non-coplanar).

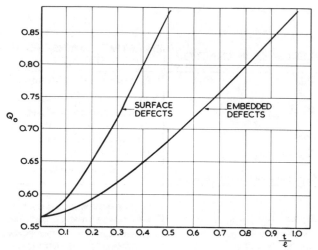

FIG. 3.   Shape factor Q_0 for different defect aspect ratios T/L.

surface effects, and they are given in the form of curves in Figs. 4–7, for embedded and surface defects, and tension and bending stress situations.

For through thickness defects the stress intensity factor is calculated from:

$$K_I = 1.25\sigma_I\sqrt{l}$$

where σ_I is the maximum tensile stress $(P_m + P_b + Q + F)$.

In curved shells there is a magnification effect on K_I for long defects related to the parameter $1/\sqrt{Re}$. This is covered in this part of the WEE/37 document by requiring the value of K_I determined as above to be multiplied by the factor $I/\sqrt{D}$ where values of D are given in Fig. 8.

For defects in stress concentration regions, either the elevated value of stress in the area containing the defect must be used or individual solutions for the particular case derived from published solutions (e.g. reference 21) or, for example, by finite element computer analysis of the particular case.

3.3. Known Defects—Above Yield Stress

This approach is to be used when the sum of $P_m + P_b + Q + F$ exceeds the yield stress. From a knowledge of the actual defect size and the applied stresses an 'effective' defect parameter $\bar{a}$ is determined. If this effective defect parameter is smaller than the 'tolerable' defect parameter, $\bar{a}_m$, then the defect is acceptable. The 'tolerable' defect parameter is determined from a

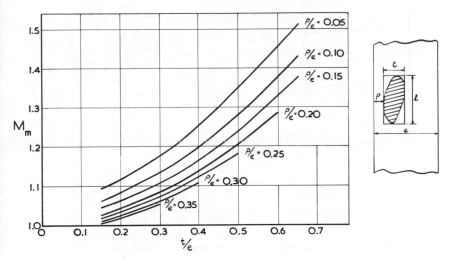

FIG. 4.   Correction factor M_m for embedded defects in tension.

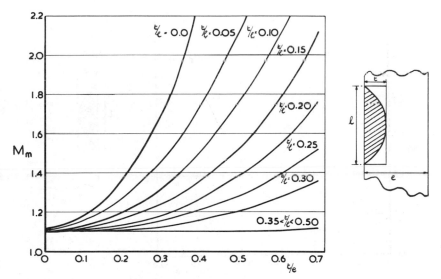

FIG. 5.   Correction factor M_m for surface defects in tension.

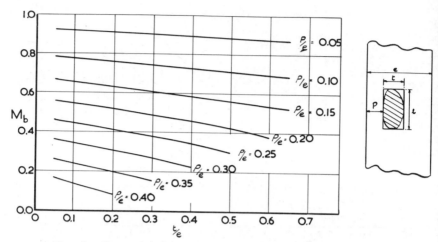

FIG. 6. Correction factor M_b for embedded defects in bending.

knowledge of the fracture toughness of the material in which the defect lies and the applied stresses. The safety factor between the tolerable and critical size of defect to cause failure is about 2 to 2·5 on average, and the 'tolerable' size is set at statistical limits equivalent to 95 % confidence that the tolerable size is less than the critical size allowing for scatter in toughness tests, etc.

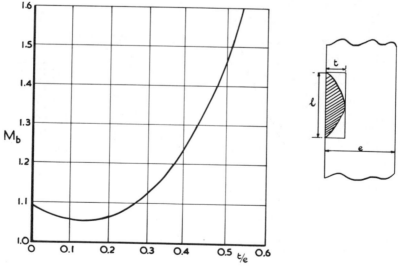

FIG. 7. Correction factor M_b for surface defects in bending.

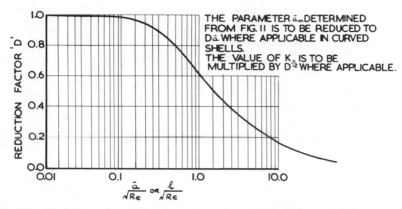

FIG. 8.   Reduction factors for long flaws in curved shells containing pressure. (Applies to any planar flaw in a sphere, and to longitudinal flaws in a cylinder.)

As in the section for below yield stresses the defect is first represented by an equivalent flaw of height t and length l. The value of the effective parameter $\bar{a}$ is then determined from Table 1 and Figs. 9 and 10 for the appropriate case. Figure 9 is used for surface defects, and Fig. 10 for embedded defects. In fact these curves are based on the same information as Figs. 4–7,[14–20] but are presented in alternative form to aid the $\bar{a}$ parameter method. Item 2 of Table 1 shows a small penalty for defects which are

TABLE 1

DEFINITIONS OF THE PARAMETER $\bar{a}$

Flaw (or Interacted Group of flaws)	Definition of $\bar{a}$
(1) Through thickness flaw as defined by Fig. 1(a)	$\bar{a} = l/2$
(2) Flaws (embedded or surface) re-categorised according as through thickness flaws due to small ligaments	$\bar{a} = (l/2) + t$
(3) Surface flaws for which $(t/e) \geq 0\cdot5$	$\bar{a} = (l/2)$
(4) Surface flaws for which $(t/e) < 0\cdot5$	See Fig. 9
(5) Embedded flaws for which $(P/t) \leq 0\cdot5$	Use flaw height $t + p$ and treat as surface flaw Fig. 9
(6) Embedded flaws for which $(P/t) > 0\cdot5$	See Fig. 10
(7) Flaw at fillet weld toe for which $(t/e) < 0\cdot5$	See Fig. 9[a]
(8) Flaw at hole for which $l \leq 0\cdot15R$	$\bar{a} = l^{a}$
(9) Flaw at hole for which $l > 0\cdot15R$	$\bar{a} + l + R^{a}$

[a] But see text for stress values

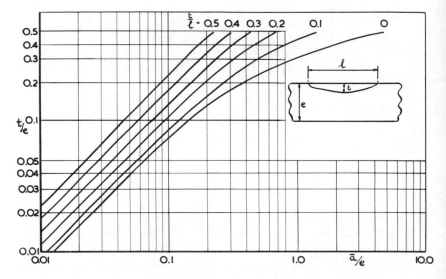

FIG. 9. Relationship between actual defect dimensions and the parameter $\bar{a}$
surface defects.

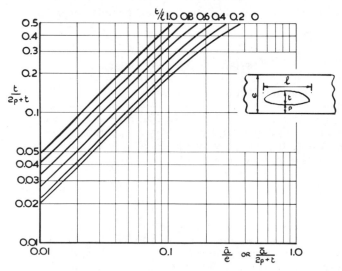

FIG. 10. Relationship between actual defect dimensions and the parameter $\bar{a}$
embedded defects.

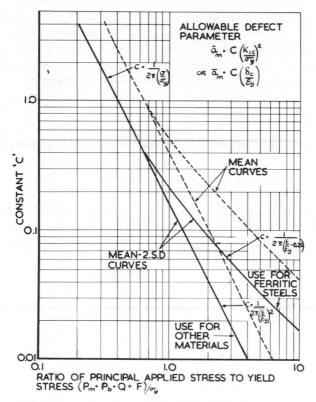

FIG. 11. Values of constant C for different loading conditions.

actually part thickness but have only a small surface ligament so that they are re-categorised as through thickness. This is because in the limit the deformation of the flaw will be controlled by its length l, but the small surface ligament will in fact be subjected to very high stress/strain conditions, and there is a risk of 'snap through' behaviour initiating a fracture in the ligament at mid-length.

The tolerable flaw parameter, $\bar{a}_m$, is determined from the fracture toughness and applied stress conditions by the relationships:

$$\bar{a}_m = C \frac{\delta_c}{\varepsilon_y}$$

where δ_c is the critical value of COD; ε_y is the yield strain, σ_y/E; and C is determined from the applied stresses using Fig. 11.

The input for the abscissa in Fig. 11 is the ratio $(P_m + P_b + Q + F)/\sigma_y$ provided yielding is contained and strains are essentially linear. In fact, if the sum of the stresses including thermal but excluding residual exceeds $2\sigma_y$, it is recommended that a full elastic–plastic stress analysis should be carried out. If the actual peak strains are known, however, the ratio $\varepsilon/\varepsilon_y$ may be used as the input to Fig. 11 to determine C. For flaws at the toe of a fillet weld or at the crotch of a pressure vessel nozzle a simplified treatment is given based upon assuming a stress concentration factor of 3 applying to the area containing a defect for a specified depth. This depth is 15 % of the thickness for defects at the toe of a fillet weld, and 15 % of the nozzle radius for a nozzle. For larger defects at a nozzle the defect length l should be increased by the nozzle radius to $(l + R)$.

The basis for Fig. 11 is the 'design curve' approach, produced by work at the Welding Institute.[6,13] This resulted originally from experimental measurements of crack opening displacement deformation against applied stress and strain for a series of large scale (1 m square) tensile tests with different notch depths. It has since been supported by analysis of service failures and of a large number of wide plate tension tests to failure, which have shown the design curve to be positioned at the mean–two standard deviation location of a plot of critical against predicted allowable flaw size.

The effect of bulging or local collapse in pressure vessels with long flaws has still to be taken into account, and the value of $\bar{a}_m$ determined as above should be reduced by the factor D from Fig. 8 to allow for this.

The defect is acceptable if $\bar{a}$ is less than $\bar{a}_m$.

3.4. Estimation of Tolerable Sizes of Planar Defect
Where it is required to lay down acceptance levels for flaws in the general situation, great emphasis has been placed by the WEE/37 and IIW documents on the need to allow for various unknown factors which may occur. Different allowable flaw sizes will be permissible at different locations in a vessel or structure depending on the applied stresses, but allowance must be made for unintentional overload or fault conditions, and variations in material properties from optimum. Attention is drawn to the effects of ovality in vessels, angular distortion, misalignment and residual stresses at welded joints, as well as inaccuracies in non-destructive testing methods. Ideally, the variability of material properties and stress about the assumed design values should be treated statistically to ensure that the standard for detection of a tolerable defect size at a required probability level can be established.

The value of the tolerable flaw parameter, $\bar{a}_m$, is determined as in Section

3.3 from the applied loading conditions and material fracture toughness. In this case, however, it is also permitted to use plane strain linear elastic fracture toughness values and either of the following relationships may be used:

$$\bar{a}_m = C\frac{(\delta_c)}{\varepsilon_y} \qquad \bar{a}_m = C\left(\frac{K_{Ic}}{\sigma_y}\right)^2$$

where C is determined from Fig. 11.

The tolerable flaw parameter may be equivalent to a number of alternative actual flaw sizes and shapes. It is therefore necessary to calculate actual allowable flaw sizes from the parameter $\bar{a}_m$ using Figs. 9 and 10, and Table 1 for each of the three categories:

(1) Through thickness flaws.
(2) Surface flaws.
(3) Embedded flaws.

In the case of long flaws in pressure vessels, the parameter $\bar{a}_m$ should be reduced by the factor D from Fig. 8 before converting into allowable flaw sizes.

In stress concentration regions the elevated stress or strain must be used in determining the value of C from Fig. 11, and likewise residual stresses must be included in the stress sum $P_m + P_b + Q + F$. If the dimensions of allowable defects calculated as above are such that re-categorisation of the flaw size from embedded to surface, or surface to through thickness would be required, the allowable size is reduced to that at which re-categorisation would not apply.

3.5. Non-planar Defects
The limits for porosity and inclusions considered acceptable in the WEE/37 and IIW documents in respect of brittle fracture failure are given in Table 2 for ferritic materials having a Charpy V-notch energy absorption of not less than 40J at the minimum service temperature. These limits also apply to other materials provided the minimum fracture toughness of the material in which the defects lie exceeds K_{Ic} 1300 Nmm$^{-3/2}$. For materials of lower toughness these defects should be assessed on the basis of their dimensions and possible interaction as planar defects. The limits take account of the possibility of the less harmful non-planar defects shielding planar defects.

3.6. Imperfect Shape
Variations from weld profile specified which would give a weld throat

TABLE 2
ACCEPTANCE LIMITS FOR NON-PLANAR DEFECTS

Type of defect	Position in thickness	Limits for brittle fracture considerations[1]	Limits for fatigue loading permitted for quality type shown					
			Q0	Q1	Q2	Q3	Q4	Q5–Q10
Slag inclusions (as-welded)	Any except at surface	Acceptable in not more than two occurrences at the same cross-section, provided the cross-section of each inclusion is less than 3 mm, and if two inclusions are present they are separated by sound metal of extent at least twice the width of the inclusion of larger cross-section	Length ≯1·5 mm	Length ≯2·5 mm	Length ≯4 mm	Length ≯10 mm	Length ≯35 mm	Continuous A–W
Slag inclusions (stress-relieved)		Limits on multiple inclusions at same cross-section apply as for brittle fracture considerations	7·5 mm	19 mm	58 mm	Continuous	Continuous	Continuous S–R
Porosity/ piping	Any except at surface	Acceptable up to 5% projected area on a radiograph, or to an extent which could, in the opinion of the Engineers, mask and prevent the detection of other more harmful defects. The maximum diameter of individual pores or pipes permitted is 25% of the plate thickness. Linear porosity is not permitted unless further checks are carried out to determine whether lack of penetration/fusion defects are also present and the flaws treated on that basis	Not acceptable	3%	3%	5%	5%	5%
			Limits and restrictions as for brittle fracture consideration apply					

dimension below that required to carry the maximum allowable design stresses are not acceptable.

Provided it can be shown that no planar defects are associated with undercut, undercut up to a depth of 1 mm or 10% of the thickness is acceptable for fracture considerations excluding fatigue, in ferritic steels with Charpy V-notch energy absorption of not less than 40J at the minimum service temperature, or other material with minimum fracture toughness K_{1c} 1300 Nmm$^{-3/2}$.

4.   DETAILED ASSESSMENT FOR FATIGUE

As for Section 3 on brittle fracture, the assessment for fatigue failure is divided into planar defects (and including non-planar surface defects) and buried non-planar defects.

The assessment methods for planar defects are based upon fracture mechanics analyses, and upon the basic crack propagation law:

$$\frac{da}{dN} = A(\Delta K)^m$$

where da/dN is the rate of crack propagation per cycle, A and m are parameters which depend on the material and applied conditions including environment and cyclic frequency, and ΔK is the range of stress intensity factor corresponding to the applied stress cycle and instantaneous fatigue crack dimensions.

The analysis for planar defects is divided into the same two categories for fatigue as for brittle fracture, namely treatment for known defects, and estimation of tolerable defects in the general case.

The basic principle of the treatment for planar defects is to estimate the total number of cycles for a crack to grow from an initial to a final size by integration of the basic crack propagation law. The final size is determined either by the crack penetrating the full thickness, or by the critical size for brittle fracture failure. The initial size is then the allowable size of flaw under the particular fatigue conditions.

For non-planar defects, the assessment is based on a large number of experimental results of fatigue tests using both natural and artificial flaws. The test results have been collected and analysed from a number of international research laboratories.

4.1. Data Required
The same terminology for stresses, defect dimensions, etc. is used in the

fatigue analysis as for the brittle fracture section. The stress to be considered is the cyclic stress range which is the sum of the cyclic components of P_m, P_b, Q and F.

In stress relieved structures the ratio R of the minimum to the maximum value of the absolute stress level $P_m + P_b + Q + F$ is required in the calculations for each cycle.

In one of the alternative methods of calculations described below it is necessary to separate the cyclic stresses into tensile and bending components. The defect dimensions and interaction effects are defined in the same way for fatigue effects as for brittle fracture.

4.2. Known Defects

Two methods are described for estimating growth of planar defects and of non-planar defects which break the surface. The first method is a general treatment which permits accurate estimates of the cyclic stress intensity factor based on precise solutions for particular defects, and uses crack propagation data constants specific to the material and conditions relevant.

The second method is a simplified procedure with an inbuilt propagation law and integration procedure, but with conservative assumptions of the stress intensity factor.

In the general procedure of the first method, the quantities A and m of the basic crack propagation law have to be agreed by all parties for the particular case, together with the threshold value of ΔK below which crack propagation does not occur. The range of stress intensity factor, ΔK, is estimated for the particular defect and stress conditions, and the growth of the flaw, da, calculated for one cycle from the crack propagation law. The stress intensity factor is then re-calculated for the increased defect size, and the whole procedure repeated until the maximum limiting flaw for failure by fracture, leakage or other modes reached. For initial surface defects, the length is kept constant until the flaw achieves a semicircular shape or breaks through the other surface. For embedded flaws the length is kept constant until the flaw achieves a circular shape, or breaks through to one surface when it becomes a surface flaw. Flaws of semicircular/circular shape are assumed to grow at constant shape until breaking through to the surfaces when they extend as through thickness defects.

For the simplified procedure of the second method, the initial defect size acceptance levels are related to quality classifications. These have been chosen to coincide with the design limits inherent to some types of welded joint detail for convenience. The classifications are given in the form of $S-N$ diagrams and are indicated in Fig. 12(a) as Q1–Q10 for the as-welded

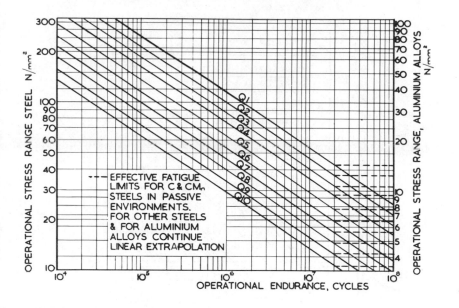

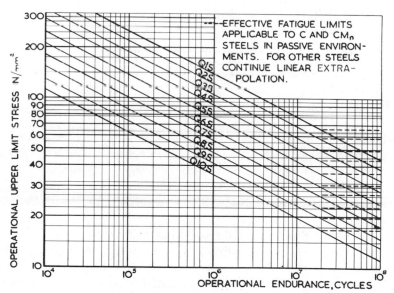

FIG. 12.   Stress/endurance relationships for quality categories as (a) welded steel and aluminium alloys, (b) stress-relieved steel.

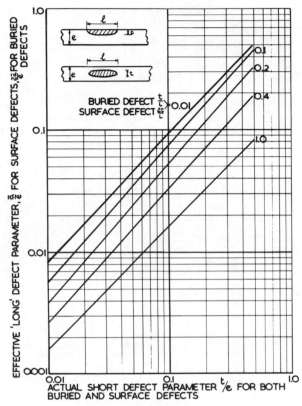

FIG. 13.   Relationship between actual 'short' defect and effective 'long' defect
dimension for fatigue loading.

condition, and in Fig. 12(b) as Q1S–Q10S for the stress-relieved condition,
for use with welded steel structures or components. The stress range
endurance lines Q1–Q5 are identical with the design limits for welded detail
categories D, E, F, F2, G in the Welding Institute Fatigue Design Rules,
also proposed for the revised Bridge Standard BS 5400, the details having a
survival probability of $97\frac{1}{2}\%$.

The user may also be given the option of using curves with a greater
certainty of $99\frac{1}{2}\%$ survival probability when considering defect acceptance.

In assessing the effect of a known defect, with initial height t and length l,
the first step is to establish an effective initial defect parameter $\bar{a}_0$ for a long
defect using Fig. 13. The curves given in Fig. 13 for different aspect ratios of
the defect, are derived from linear elastic stress intensity factor solutions for

surface and buried flaws of elliptical profile, taking into account the subsequent shape during growth by fatigue.

It is then necessary to define a limiting size of effective 'long' defect to which growth is acceptable. This may be determined by consideration of the tolerable defect size for unstable fracture, using the methods described earlier in this chapter, or, alternatively, by other service requirements such as limiting the defect height to be less than a proportion of the thickness. This limiting defect height parameter is termed $\bar{a}_m$.

A series of curves is given which effectively presents the size of initial defect which will grow to a limiting condition completely through the plate thickness, with a constant amplitude stress range, in 10^5 cycles. The choice of 10^5 cycles is an arbitrary index, which does not affect the overall fatigue calculations as will be seen later. Examples of these curves are given in the following figures:

Fig. 14(a) buried or embedded defects;
Fig. 14(b) surface defects;
Fig. 15(a) defects at toe of 5 mm leg fillet;
Fig. 15(b) defects at toe of 8 mm leg fillet; and
Fig. 15(c) defects at toe of 12 mm leg fillet.

The procedure to estimate the quality classification representing the effect of a known defect, and hence the life for given applied loading, is as follows. From the appropriate figure for the defect position (Figs. 14 and 15), the figures are first entered at $\bar{a}_0$ on the ordinate axis, and a value of the stress parameter, S, estimated for the thickness and other relevant geometry. This value is termed S_0. Using the same figure, the process is then repeated entering with $\bar{a}_m$ on the ordinate axis to estimate a corresponding value of S for the thickness and geometry, termed S_m.

The two values S_0 and S_m are then used to calculate a value of S given by:

$$S = (S_0^4 - S_m^4)^{1/4}$$

The quality classification for the defect under consideration is that next below S in Table 3.

The procedure described above, in effect, assesses the stress range for a defect $\bar{a}_0$ to grow full plate thickness in 10^5 cycles, and the stress range for a defect $\bar{a}_m$ to grow to full plate thickness, and calculates an effective stress range for the defect to grow from $\bar{a}_0$ to $\bar{a}_m$ in 10^5 cycles. The use of 10^5 cycles is an arbitrary index, used purely to calculate the quality classification. For aluminium alloys the same procedure outlined above can be used, provided the stress values used are divided by three.

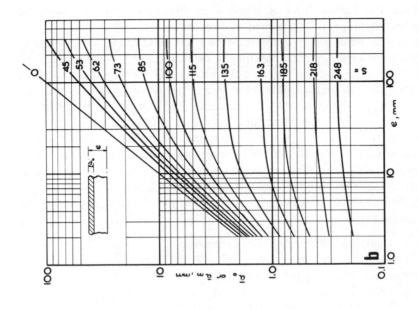

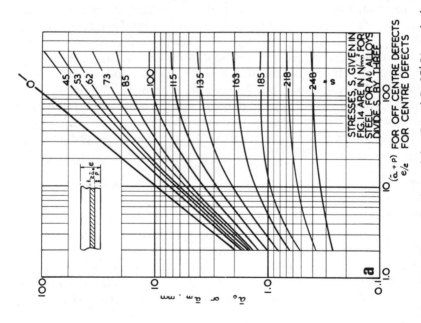

Fig. 14. Curves for determining S_o and S_m $97\frac{1}{2}\%$ survival probability. (a) Buried defects; (b) surface defects, flat plate.

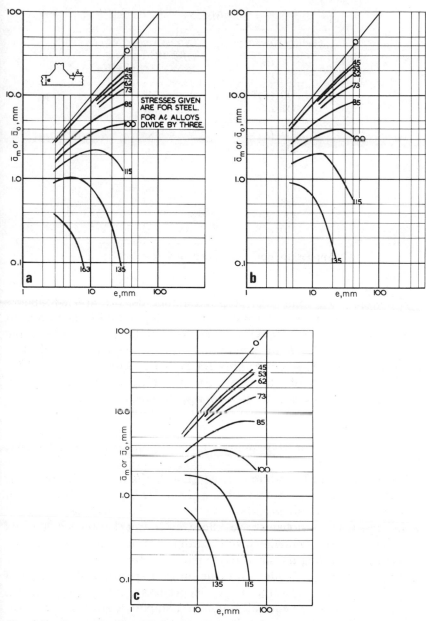

FIG. 15. Curves for determining S_o and S_m for defects at the toes of fillet welds, for $97\frac{1}{2}\%$ survival probability. Leg length fillet weld: (a) 5 mm, (b) 8 mm, (c) 12 mm.

TABLE 3

Quality category	Stress range, S, for 10^5 cycles in steel (N/mm^2)
Q0 $(S)^a$	> 248
Q1 (S)	248
Q2 (S)	218
Q3 (S)	185
Q4 (S)	163
Q5 (S)	135
Q6 (S)	115
Q7 (S)	100
Q8 (S)	85
Q9 (S)	73
Q10 (S)	62

^{a} Applies to both as-welded and stress relieved conditions.

If the applied loading is at constant amplitude, the 'applied' quality classification can be determined directly from Fig. 12(a) for the as-welded condition, or Fig. 12(b) for the stress-relieved condition. In the case of the stress-relieved condition, the stress range to be considered is an equivalent pulsating tension stress range. Thus for this case, for cycles with stresses passing through zero only the tensile part of the stress range is considered, and for cycles with a positive stress range S_R only, S is taken as:

$$S = S_R(0.25R + 1)$$

where R is the ratio of minimum to maximum absolute stress level ($P_m + P_b + Q + F$).

For the as-welded condition, at constant amplitude stress range, the 'applied' quality classification is obtained directly from Fig. 12(a) without adjustment for mean stress and stress ratio. This is because the presence of yield stress tensile residual stresses effectively means that the actual stress range at the tip of an initial flaw will be from yield stress down over the full range, thus adjusting the stress range into the tension region.

Where the applied loading is made up of a spectrum of loadings of variable amplitude, consisting of n_1 cycles of stress range S_1, n_2 cycles at S_2, n_i cycles at S_i, etc., the equivalent constant amplitude stress range at 10^5 cycles is calculated. This procedure differs for the as-welded and stress-relieved conditions because of the effect of residual stress and mean stress on fatigue crack propagation. This effect results in the slopes of the quality classification lines in Fig. 12(a) being $-\frac{1}{3}$ for the as-welded condition and in Fig. 12(b) being $-\frac{1}{4}$ for the stress-relieved condition, the corresponding lines in the two figures passing through the same stress range at 10^5 cycles, however. The equivalent constant amplitude stress range at 10^5 cycles is calculated as follows:

As-welded

$$S = \left(\frac{\sum n_i S_i^3}{10^5}\right)^{1/3}$$

Stress-relieved

$$S = \left(\frac{\sum n_i S_i^4}{10^5}\right)^{1/4}$$

The acceptability of a known defect is determined by comparing the quality classification of the defect effect with the applied loading conditions. If the quality classification of the defect is less severe than the applied loading the defect is acceptable.

It will be found from operation of the procedures described above that in some cases only very small defect sizes are acceptable. Indeed the analyses may predict acceptable defect size limits below the sizes which could be detected by realistic non-destructive testing methods. It will also be found that the allowable sizes of defects at the toes of fillet welds are undetectable when the operating stress range and design life are up to the maximum permitted in the fatigue design rules for such details. This situation arises because the design rules are based upon defect-free welds and the presence of any defects will reduce the fatigue life. The implication of finding the calculations to show exceedingly small allowable defect sizes is that the design conditions should be re-checked to find methods of designing round the problem by avoiding the presence of such critical details in a region subject to the stress level and life requirements which gave this result.

4.3. Estimation of Tolerable Sizes of Planar Defect

The estimation of tolerable sizes of planar defects is carried out using the same techniques as described above but in reverse order, and using Figs. 12 to 15 inclusive. This procedure will result in calculating the allowable 'long' defect, $\bar{a}$, and it will be necessary to consider a number of possible defect shapes and suspect regions for 'short' defects using Fig. 13. The 'long' defect parameter, $\bar{a}_0$, will give the limiting condition, however. In estimating tolerable defect sizes the most severe combination of defect orientation and position, and the most stringent combinations of cyclic loading should be considered.

4.4. Non-planar Defects

The allowable defects given in the WEE/37 approach for non-planar defects under fatigue conditions are based upon a widespread survey of experimental data from a number of laboratories.[22,23] The requirements are set out in Table 2.

The requirement for porosity is based for the lower categories on a limiting condition to ensure that the porosity does not cloud or shield the detection of more harmful defects. The experimental work showed that the porosity itself could be considerably denser for the lower categories before having harmful effects but it was considered by the Committees that such porosity would make interpretation of non-destructing testing methods unsatisfactory.

The requirements for slag inclusions are based upon length of inclusions on the assumption that the height of the inclusions would be typical and inherent to the nature of the defect. The requirements for slag inclusions take into account the possible effects of different positions within the thickness provided the inclusions do not break surface.

5.   OTHER MODES OF FAILURE

The other modes of failure considered in the WEE/37 document are as follows:

(1)   Yielding due to overloading of remaining cross-section.
(2)   Leakage in pressure liquid or vacuum containing equipment.
(3)   Environmental effects.
(4)   Instability.
(5)   Creep.

The treatment for each of these modes is given only in general terms but the requirements are summarised below.

5.1. Yielding Due to Overloading of Remaining Cross-section

The criteria given in the document are intended only for small structural sections where the sizes of defects present may form a significant part of the total cross-section. In such cases the requirement is that the nominal stress on the reduced area of the complete cross-section should not exceed 95 % of the specified minimum yield stress and the net shear stress should not exceed 45 % of the minimum specified yield stress. The document excludes this type of simple analysis for defects in pressure vessels or any sections subject to complex loading without a detailed analysis of the stress including the effect of load re-distribution.

5.2. Leakage in Pressure Liquid or Vacuum Containing Equipment

This section of the document provides that no defects are acceptable which provide a path from interior to exterior of such equipment or which will grow to provide such a path during service. Leakage paths detected during proof testing rather than by non-destructive testing are anticipated and remedial action to repair such defects is to be taken by agreement.

5.3. Environmental Effects

General guidance is given on three types of environmental effect, namely corrosion/erosion, corrosion fatigue and stress corrosion. In the cases of corrosion/erosion and stress corrosion it is pointed out that these types of failure are influenced mainly by the choice of materials suitable for the particular environment. Mention is made of the effect of crevices or exposed defects at the surface on both corrosion and stress corrosion and exposed defects should be excluded unless estimates can be made at the rate of growth of defects in service.

The requirements for corrosion fatigue are left general that when fatigue loading occurs in an aggressive environment it is necessary for it to be demonstrated by reference to previous experience or specific tests that the stresses can be tolerated for the required lifetime of the structure.

5.4. Instability (Buckling)

Again only general guidance is given on this subject to draw attention to the possible effect of flaws on buckling behaviour. This is unlikely to be significant in pressure vessel applications except for some complex geometry joint areas near to end closures, or to vessels subject to external pressure. In such cases consideration should be given to the effect of any defects on the second moment of area of the affected cross-section when welded joints containing flaws may lie in zones of compression.

5.5. Creep

Guidance is given on non-planar defects to the effect that they should not exceed levels equivalent to those shown in Table 2 for Category E. For planar defects, assessment may be made by calculations based on the net section stress and creep or creep rupture stress data provided it is shown that the material is not sensitive to failure by macroscopic crack propagation. Where no such data or experience are available, planar defects under creep environments should be considered unacceptable. It is also recommended that periodic inspections should be carried out in such cases. Attention is drawn to the possible interaction between creep and fatigue but no definitive guidance is given.

REFERENCES

1. BRITISH STANDARDS INSTITUTION. Draft Standard Rules for the Derivation of Acceptance Levels for Defects in Welds, Technical Committee WEE/37, 1975, Draft for Comment Document 75/77081 DC.†
2. INTERNATIONAL INSTITUTE OF WELDING. Report of Commission X Working Group on the Significance of Defects, *Welding in the World*, 1975, **1/2**, 29–46.
3. *Proc. of the First Conference on the Significance of Defects in Welds*, held in Feb. 1967, Published 1968, The Welding Institute, London.
4. *Proc. of the Second Conference on the Significance of Defects in Welds*, held in May 1968, Published 1969, The Welding Institute, London.
5. BURDEKIN, F. M., HARRISON, J. D. and YOUNG, J. G., *Proc. of the First Conference on the Significance of Defects in Welds*, 1968, The Welding Institute, London, 63–74.
6. HARRISON, J. D., BURDEKIN, F. M. and YOUNG, J. G., *Proc. of the Second Conference on the Significance of Defects in Welds*, 1969, The Welding Institute, London, 65–79.
7. *ASME Boiler and Pressure Vessel Code*, Section III, Nuclear vessels.
8. *ASTM STP* 381, 1964, Fracture toughness testing and its applications, Philadelphia.
9. *ASTM STP* 410, 1967, Plain strain fracture toughness testing, Philadelphia.
10. BRITISH STANDARDS INSTITUTION. Methods for plane strain fracture toughness (K_{Ic}) testing, BS 5447:1977, London.
11. WELLS, A. A. Symposium on Crack Propagation, 1961, Cranfield College of Aeronautics, Cranfield, 210–30.
12. BRITISH STANDARDS INSTITUTION. Methods for Crack Opening Displacement (COD) testing, 1972, Draft for Development DD19.
13. BURDEKIN, F. M. and Dawes, M. G. Practical Application of Fracture Mechanics to Pressure Vessel Technology, 1971, I.Mech.E., London, 28–37.
14. KOBAYASHI, A. S., ZIV, M. and HALL, L. R. *Int. J. Fracture Mechanics*, 1965, **1**(2), 81–95.
15. SHAH, R. C. and KOBAYASHI, A. S., *J. Eng. Fracture Mechanics*, 1971, **3**(1), 71–96.
16. SHAH, R. C. and KOBAYASHI, A. S. *Int. J. Fracture Mechanics*, 1973, **9**(2), 133–46.

† Expected to be re-issued after comment and amendment as a BSI Guidance Document.

17. SMITH, F. W. and |ALAVI,| M. J. First International Conference on Pressure Vessel Technology, Oct. 1969, Delft, Part 2, 793–800.
18. SMITH, F. W. The Boeing Co. Ltd, Structural Development Research Memo. No. 17, 1966.
19. THRESHER, R. W. Ph.D. Thesis, Aug. 1970, Colorado State University.
20. THRESHER, R. W. *Trans. ASME, J. Appl. Mech.*, Mar. 1972, **39**, 195–200.
21. ROOKE, D. P. and Cartwright, D. J. *Compendium of stress intensity factors*, HMSO, 1976.
22. HARRISON, J. D. *J. Metal Construction*, 1972, **4**(3), 99–103.
23. HARRISON, J. D. *J. Metal Construction*, 1972, **4**(7), 262–8.

Chapter 4

THE CEGB TWO-CRITERIA PROPOSAL

B. J. L. Darlaston

Berkeley Nuclear Laboratories, Berkeley, UK

SUMMARY

The development of the CEGB failure assessment route by Harrison, Milne and Loosemore is described. The failure assessment diagram which is used as the basis for this approach extends linear elastic fracture mechanics principles into the post-yield regime through a two-criteria approach. This approach assumes that a failure occurs when the applied load reaches the lower of either a load to cause the brittle failure based on linear elastic fracture mechanics, or a collapse load dependent upon a flow stress and the structural gravity. The transition region is described by a strip yield model. Examples of the use of the failure assessment diagram are given.

The failure assessment diagram is then presented as part of the failure assessment route with a code format similar to that of ASME XI Appendix A.

1. INTRODUCTION

In the past decade increasing attention has been paid to the subject of fracture mechanics, and more recently to the particular aspect of post-yield fracture mechanics. In the field of structural analysis, attention has in the main been focused on LEFM considerations, because of either the difficulty in applying sophisticated PYFM methods or simply because LEFM methods provided adequate assessments. With the wider use of fracture mechanics and the need to apply the techniques to higher toughness structures, attention began to turn to PYFM assessment methods. One such method has been developed by Harrison, Milne and Loosemore[1] and is known as the CEGB failure assessment route. In developing the failure assessment route, the overall objective was to present a method based on established and well understood techniques brought together in a code format which was familiar to the practising engineer. There was no question of developing a more advanced PYFM methodology, but simply to bring together existing knowledge.

2. THE FAILURE ASSESSMENT DIAGRAM (FAD)

In the proposed method the avoidance of failure is assessed by reference to the failure assessment diagram (not to be confused with fracture analysis diagrams proposed by other workers) shown in Fig. 1. If the point defined by K_r, S_r falls inside the envelope of the assessment curve, failure will be avoided.

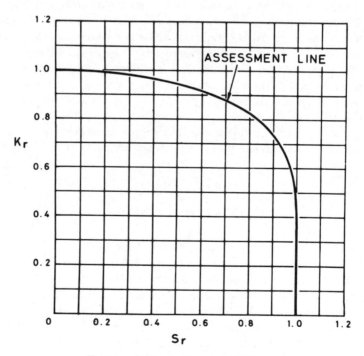

FIG. 1.   Failure assessment diagram.

The definitions of K_r and S_r are as follows:

$$K_r = \frac{\text{Stress intensity factor}}{\text{Fracture toughness}} = \frac{K_I}{K_{Ic}}$$

$$S_r = \frac{\text{Applied load}}{\text{Collapsed load}}$$

The stress intensity factor is based on LEFM principles, and K_{Ic} is the plain strain fracture toughness. The applied load and the collapsed load can be defined in terms of bending moment, pressure or simply dead load.

3. BASIS OF THE FAD

Dowling and Townley[2] postulated a two-criteria approach to failure assessment of structures containing defects. This assumes that a failure occurs when the applied load reaches the lower of either a load to cause the brittle failure based on linear elastic fracture mechanics, or a collapse load dependent upon a flow stress and the structural geometry. A sizeable transition exists between the two criteria, and it has been shown that this can be described by an equation developed from the Bilby Cottrell Swinden strip yield model.[3] The expression so derived adequately predicts the behaviour of crack structures of simple and complex geometry. A universal failure curve based on the two-criteria approach was subsequently proposed As the development and validation of this two-criteria approach forms the basis of the failure assessment diagram and, as such, of the CEGB assessment route the postulation will be examined in more detail.

3.1. PYFM Methods of Analysis

When the remote gross tensile stress is less than about 0·6 of the uniaxial stress of the material, stress intensity approach to brittle fracture is a good way of describing the effect of cracks in the structure. At higher stresses, large plastic zones are formed at the crack tip, and it is under these conditions that various forms of post-yield fracture mechanics are used to give better appreciation of failure stresses and defect sizes. PYFM methods are well covered in the literature and discussed elsewhere in this book, and it is not proposed to cover them here, save to mention the approach used in the development of the two-criteria method.

In linear elastic mechanics theory[4,5] it is implicitly assumed that the material ahead of the crack can support infinite stress. Any post-yield theory must include some representation of plastic deformation at the crack tip. The work of Bilby, Cottrell and Swinden, who presented a strip yield model, was taken in the form modified by Heald, Spink and Worthington.[6] The resulting equation is of the form:

$$\sigma_f = \sigma_u \frac{2}{\pi} \cos^{-1} \left[\exp - \frac{\pi^2}{8} \left(\frac{K_{Ic}}{y\sigma_u} a^{1/2} \right)^2 \right] \tag{1}$$

3.2. Plastic Collapse and Limit Analysis

For tough structures, the net section stress will approach yield, at which point gross deformation can occur. For plastic collapse the structure must become a mechanism. In the case of an elastic, perfectly plastic material, once plastic hinges have formed gross deformation will occur with no increase of external load. If a crack is situated in a region of high stress it can cause the plastic collapse load of the structure to be altered. In practice materials work-harden and the load corresponding to the onset of gross deformation is different from the load corresponding to total failure or ultimate collapse. As will be discussed later, it is necessary to make assumptions regarding the flow stress to be used in such calculations.

3.3. Interaction of LEFM and Limit Analysis

As already stated the two-criteria approach to the evaluation of defects in structures states that failure occurs when the loading system reaches the lower of either the magnitude calculated to cause failure by linear elastic fracture mechanics theory, or a magnitude sufficient to cause ultimate collapse. Under the extreme conditions of a very brittle and very tough material, both these criteria are well validated experimentally. The transition region between the two is the aspect on which attention will be focused, but first of all it is necessary to consider the two bounds.

For an arbitrary cracked body from LEFM theory

$$K_{Ic} = C\sigma_0 (\pi C)^{1/2}$$

i.e.

$$\sigma_{of} = \frac{A(C/w) K_{Ic}}{(\pi C)^{1/2}} \tag{2}$$

and from limit analysis, the ultimate collapse stress can be stated as

$$\sigma_{ou} = B(C/W)\sigma_u \tag{3}$$

where $A(C/W)$ is a function of crack length to wall thickness ratio; K_{Ic} is the material fracture toughness; $B(C/W)$ is a function of C/W; σ_u is the ultimate stress; σ_{of} is failure stress from LEFM theory; and σ_{ou} is the ultimate collapse stress of the structure.

Noting that the ultimate collapse stress of the uncracked body can be stated as $\sigma'_{ou} = D\sigma_u$, where D is a constant dependent upon geometry, eqns. (2) and (3) can be re-written to give:

LEFM

$$P_f^\varphi = \frac{\sigma_{of}}{\sigma'_{ou}} = \frac{A(C/W)}{D} \frac{K_{1c}}{\sigma_u} \frac{1}{(\pi C)^{1/2}} \left(= \infty \quad \text{for } \frac{C}{W} = 0 \right)$$

Ultimate collapse

$$P_u^\varphi = \frac{\sigma_{ou}}{\sigma'_{ou}} = \frac{B(C/W)}{D} \left(= 1 \text{ for } \frac{C}{W} = 0 \right)$$

where P^φ is a non-dimensional failure parameter.

Several fracture mechanics–limit analysis diagrams can be produced from eqns. (4) and (5) as shown in Fig. 2.

The diagrams in Fig. 2 were used extensively by Dowling and Townley, to demonstrate the validity of their two-criteria approach.

3.4. The Two-Criteria Universal Failure Curve

Dowling and Townley analysed various structural geometries and materials and concluded that as an assessment approach the diagrams shown in Fig. 2 each had their drawbacks. In addition, a clearer definition of the transition region was required.

Normalisation of the data was achieved using the formulation of Heald *et al.*, eqn. (1), which includes a description of the transition zone. Dowling and Townley used this equation to describe the behaviour of simple specimens in brittle fracture, gross yielding and the transition region. To describe the failure of more complex structures they re-wrote the formula of eqn. (1) as

$$\frac{L_f}{L_u} = \frac{2}{\pi} \cos^{-1} \left[\exp - \left(\frac{\pi^2}{8} \frac{L_K^2}{L_u^2} \right) \right] \tag{4}$$

where L_f = a failure parameter; L_u = limit load; and L_K = failure according to LEFM. This is shown schematically in Fig. 3.

Results of a large number of analyses have been plotted on such a diagram by Darlaston *et al.*[7] and are shown in Fig. 4.

There is very good agreement between the curves and the experimental points bearing in mind the wide variety of geometries, materials and

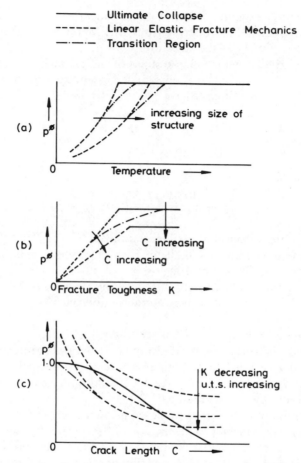

FIG. 2.   Fracture mechanics/limit analysis diagrams.

toughness levels. Although the approach has an adequate theoretical basis for simple specimens, it is empirical for complex structures.

There are a number of minor difficulties in the use of this method of failure assessment, one of which is basing a limit analysis on the ultimate stress. This is an empirical misprocedure, and the calculation of ultimate collapse loads needs to be studied more rigorously.

3.5. Defining the Failure Assessment Diagram (FAD)

In finally producing the failure assessment diagram, Harrison *et al.* concluded that eqn. (4), in terms of load with crack tip and net section events

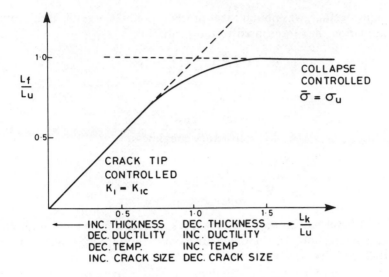

FIG. 3.   Failure assessment—two-criteria approach.

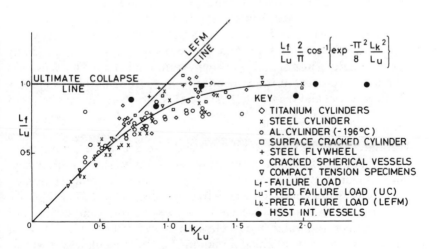

FIG. 4.   Validation of two-criteria approach.

brought together, was not an ideal presentation. As a result of algebraic manipulation they transposed eqn. (4) into

$$K_r = S_r \left\{ \frac{8}{\pi^2} \ln \sec\left(\frac{\pi}{2} S_r\right) \right\}^{-1/2} \tag{5}$$

where

$$K_r = \frac{\text{stress intensity factor}}{\text{fracture toughness}}$$

and

$$S_r = \frac{\text{applied load}}{\text{collapse load}}$$

This curve effectively separated the crack tip events from those associated with the net section. The failure assessment diagram, Fig. 1, allows a very rapid analysis by merely plotting K_r and S_r and examining where the point lies in relation to the curve. If the point is inside the envelope failure will be avoided.

4. APPLICATION OF THE FAD

4.1. Pressure Vessel Analysis
One example used to demonstrate the use of the failure assessment diagram is vessel no. 5 from the United States Atomic Energy Commission Programme of Heavy Section Steel Tests (HSST Programme). The vessel consisted of a 991 mm OD, 152 mm thick-walled cylinder with a radially attached nozzle 530 mm OD, 152 mm wall thickness. The flaw was 31·75 mm deep into the inside corner of the intersection of the axial planes of the main vessel and the nozzle. The vessel is shown in Fig. 5. The vessel failed at a pressure of 183·5 MNm^{-2} at 88°C, and it is under these conditions that the vessel will be assessed.

The material of the vessel was A533B with average tensile properties at 88°C of 425 and 553 MNm^{-2} for yield and ultimate stress respectively, obtained in pre-test information.[8] From 22-mm-thick compact tension specimens using the Witt technique[9] a fracture toughness of 270 MNm$^{-3/2}$ was determined. As no details have been given of this analysis it is not known whether the value used refers to a maximum load apparent toughness, or the value at crack initiation. For this calculation it will be taken as a true minimum.

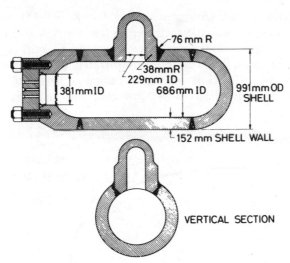

FIG. 5.   HSST intermediate vessel with test nozzle.

An elastic–plastic finite element calculation on the nozzle geometry was performed by Merkle[8] and this gave a collapse pressure of 224 MNm^{-2} for the uncracked intersection. Assuming, conservatively, that the presence of a flaw reduces this pressure linearly with increasing flaw depth then

$$S_r - \frac{L}{L_{uc}(1 - a/t)}$$

where L is the applied pressure; L_{uc} is the collapse pressure for the unflawed intersection (224 MNm^{-2}); a is the flaw depth; and t is the width of the structure at the intersection (241 mm). The resulting value of S_r was 0·94.

From Rashid and Gilman[10]

$$K_I = \sigma_h \sqrt{\pi a} F\left(\frac{a}{R_i}\right)$$

where σ_h is the circumferential stress in the vessel and is given by

$$\sigma_h = \frac{L R_m}{t_v}$$

where R_m is the mean radius of the vessel; t_v is the wall thickness of the vessel; and $F(a/R_i)$ is 1·6 for the flaw depth/radius considered. Taking K_{Ic} as 270 MNm$^{-3/2}$ the resulting value of K_r is 0·95.

Plotting the assessment point $(S_r = 0.94, K_r = 0.95)$ on the failure assessment diagram it is shown in Fig. 6 that the flaw is unacceptable.

The advantages of the approach presented through the failure assessment diagram are readily demonstrated by this pressure vessel example. In any assessment method it is essential to be able to define the way in which assumptions made influence the final result, or which

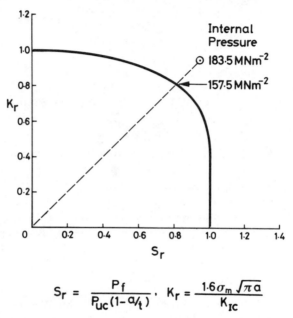

$$S_r = \frac{P_f}{P_{uc}(1-a/t)}, \quad K_r = \frac{1.6\sigma_m\sqrt{\pi a}}{K_{IC}}$$

FIG. 6.   Failure assessment diagram: effect of pressure.

parameters have the greatest effect on safety margins. A great advantage of the failure assessment diagram procedure is that this is easily done and the results can be graphically presented. Three examples will be used to demonstrate the point.

4.1.1. *Applied Pressure*
Both S_r and K_r are linearly dependent on pressure, and reduced to zero at zero applied pressure. Referring to Fig. 6 it can be seen that failure will be avoided at the intersect of the line constructed through the origin and the assessment point S_r, K_r, with the assessment line. This occurs at a pressure of $157.5\,\text{MNm}^{-2}$ which is 0.86 of the measured failure pressure. This

indicates that a safety factor of 0·6 on applied pressure has built into the safety analysis for this particular situation. This factor arises from the convention of a pessimistic approach to fracture mechanics analysis, and emphasises the fact that the proposed failure assessment diagram is used to avoid failure rather than to predict failure.

4.1.2. Fracture Toughness
In many analyses the value of fracture toughness is not available, and estimates have to be made. The failure assessment diagram readily shows

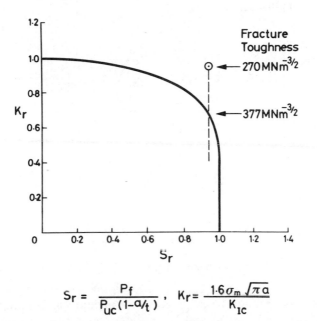

$$S_r = \frac{P_f}{P_{uc}(1 - a/t)}, \quad K_r = \frac{1\cdot 6\sigma_m \sqrt{\pi a}}{K_{IC}}$$

FIG. 7. Failure assessment diagram: effect of fracture toughness.

the effects of error in such estimates. S_r is clearly not dependent on fracture toughness whereas K_r is inversely proportional to fracture toughness. The sensitivity of the assessment point to fracture toughness can therefore be readily assessed by a vertical line through the assessment point, as shown in Fig. 7.

In this particular example it is shown that if all the other assumptions and calculations were exact, then a fracture toughness in excess of 377 MNm^{-2} would have prevented failure of the vessel at the pressure of 183·5 MNm^{-2}.

4.1.3. *Allowable Defect Size*

The effect of varying the defect depth only is shown in Fig. 8. This figure illustrates the fact that the flaw depth would have to be reduced to 20 mm in order to avoid failure.

This particular part of the analysis illustrates a very important point with regard to safety factors and code requirements. As stated, the flaw depth

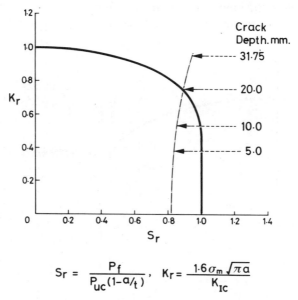

$$S_r = \frac{P_f}{P_{uc}(1-a/t)}, \quad K_r = \frac{1.6\,\sigma_m\,\sqrt{\pi a}}{K_{IC}}$$

FIG. 8.   Failure assessment diagram: effect of allowable defect size.

predicted to avoid failure is seen to be 20 mm and this can be compared with the LEFM prediction of 34 mm. Let us now consider the question of code requirements with specific reference to ASME Section II Appendix A of the Pressure Vessel and Boiler Code. This code requires a factor of 10 on the LEFM calculation critical flaw size to define the maximum allowable flaw size. Under elastic conditions this provides a safety factor of approximately three on the applied pressure. With reference to Fig. 7 it can be seen that where there is large-scale plasticity the safety factor is much reduced. The safety factor is defined in this instance by drawing a straight line joining the assessment point to the origin and intercepting the assessment line. The required safety factor is in the ratio of the distance from the assessment point from the origin to the distance of the intercept from the origin. For an applied pressure of $183 \cdot 5\,\mathrm{MNm}^{-2}$ it can be seen that this factor is a maximum in the absence of a flaw, i.e. when K_r is zero and is only $1 \cdot 22$. This value of $1 \cdot 22$ represents a considerable reduction in the safety factor compared with the LEFM value of 3.

4.2. Reactor Diagrid Analysis

This problem is somewhat different in nature in so far that pressure loading is low and consideration is given to residual stresses arising from fabrication.

The diagrid is that part of the reactor structure which supports the core, and is shown schematically in Fig. 9.

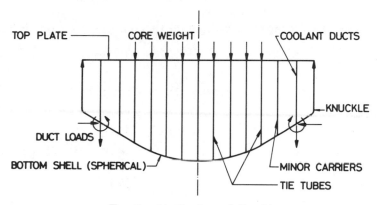

FIG. 9. Idealisation of diagrid.

The analysis is carried out in several parts. For the three critical areas of the diagrid as indicated in Fig. 9, i.e. the top plate, bottom shell and the knuckle, critical defect sizes have been calculated for plastic collapse and LEFM fracture modes. The top plate has then been assessed in detail using the failure assessment diagram.

The structure is subjected to a core weight of 259 tonnes and an internal pressure of $860\,\mathrm{kNm}^{-2}$. A finite element analysis was carried out and for the combined core weight and pressure the results obtained were as shown in Table 1, which also gives details of the appropriate section thickness.

It should be noted in the situations where a bending stress is present it is assumed that the defect runs perpendicular to the plane of bending as this will have the greatest effect on the limit load.

The material is assumed to be rigid plastic with a yield/flow stress of $180\,\mathrm{MNm}^{-2}$ and the calculations have been repeated using a flow stress of

B. J. L. DARLASTON

TABLE 1

Position	Wall thickness (mm)	Maximum membrane stress (MN/m^2)	Maximum bending stress (MN/m^2)
Top plate	10	43	—
Bottom shell	40	25	150
Knuckle	40	25	70

$315\,MNm^{-2}$, this being the half ($\sigma_u + \sigma_y$) stress, corresponding to an ultimate stress of $450\,MNm^{-2}$. For the purpose of the limit analysis the elastic stresses had to be converted to membrane force/unit length and bending moment/unit length of shell.

The results for the top plate, bottom shell and knuckle are shown in Table 2.

TABLE 2
DIAGRID—CRITICAL CRACK DEPTH (mm) LIMIT ANALYSIS

Position	$\sigma_R = 0$ $\bar{\sigma} = 180\,MN/m^2$	σ_R type 1 $\bar{\sigma} = 315\,MN/m^2$	σ_R type 2 $\bar{\sigma} = 315\,MN/m^2$	$\sigma_R = 0$ $\bar{\sigma} = 315\,MN/m^2$
Top plate	7·5	3·0	6·8	8·6
Botton shell	27·4	6·0	8·9	32·8
Knuckle	31·4	12·2	16·1	35·1

It has been suggested that residual stresses may have a significant effect on the local collapse load of the structure due to the possibility of local yielding and fracture occurring before general yielding of the structure. If residual stresses are included as an equivalent mechanical load in the limit load calculations, this results in a rather pessimistic view of the critical defect size. However, because of these uncertainties the critical defect sizes for the bottom shell and knuckle region have been recalculated with yield point ($180\,MNm^{-2}$) residual stresses incorporated as mechanical loads. Two ways of representing the residual stresses have been considered. The first incorporates a mechanical load that is constant with crack penetration; the second approach incorporates the equivalent mechanical load as a function of crack penetration to represent the relaxation residual stress as the crack progresses through the wall. The validity of either of these approaches is dependent on the nature of the residual stress in the structure as a whole. Table 2 shows the result of this analysis for type 1 residual stress systems that do not relax with crack penetration, and type 2 thermal and residual stresses that do relax as the crack propagates.

For the LEFM analysis the relevant material parameter is the value of fracture toughness, and for this a value of $80 \, MNm^{-3/2}$ was used. As the question of residual stresses is far from being resolved, the two extremes have been considered, namely fully stress relieved and yield stress levels. The yield stress values and the geometries assumed are the same as those taken for the limit load analysis. The stress intensity calibration factors have been taken from Rooke and Cartwright[11] for infinitely long defects or from the methods given in ASME Section II. The results for the top plate, bottom shell and knuckle regions for the fully stress relieved and the residual yield stress level conditions are given in Table 3.

TABLE 3

DIAGRID—CRITICAL   CRACK   DEPTH   (mm)   (LEFM ANALYSIS)

Position	$\sigma_R = 0$	$\sigma_R = 180 \, MNm^{-2}$
Top plate	>7	5
Bottom shell	24	13·6
Knuckle	N/F	33

It should be noted that in the case of the stress relieved condition for the knuckle that with the fracture toughness of $80 \, MNm^{-3/2}$ that failure would not occur.

Using the failure assessment diagram a more detailed examination was made of the top plate. Four situations were considered:

(a)   stress relieved with a flow stress half $(\sigma_y + \sigma_u) = 315 \, MNm^{-2}$;
(b)   stress relieved with a flow stress $\sigma_y = 180 \, MNm^{-2}$;
(c)   assuming a residual stress of $180 \, MNm^{-2}$ acting only on the LEFM component (K_r) with a flow stress of $180 \, MNm^{-2}$; and
(d)   assuming a residual stress of $180 \, MNm^{-2}$ acting on both the LEFM component (K_r) and the limit load component (S_r) with a flow stress of $315 \, MNm^{-2}$.

The results are shown in Fig. 10.

The results show that for a stress relieved structure very deep defects are required for failure to occur. The need to clarify the situation regarding the effect of residual stresses is shown clearly in comparing cases (c) and (d). In

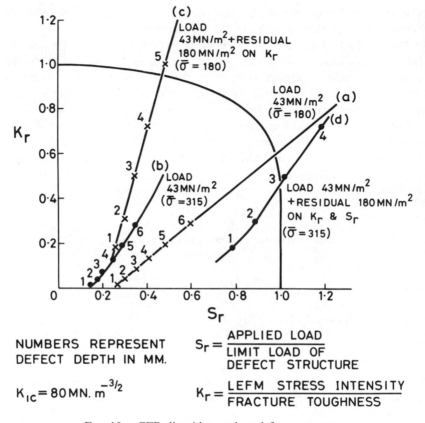

FIG. 10.   CFR diagrid top plate defect assessment.

case (c) the residual stress is considered to be an equivalent mechanical load acting on the LEFM component, and this gave a defect depth of 4·8 mm. In case (d) the residual stress was taken into account on the LEFM component and on the limit load, and this reduced the defect size to 2·9 mm. The general question of accommodating thermal and residual stresses in the failure assessment diagram approach is the subject of further research.

5.   THE FAILURE ASSESSMENT ROUTE

The failure assessment diagram is the corner-stone of the failure assessment route. Harrison *et al.* based the procedures of the failure assessment route on the format of the ASME Section XI code. This provided a document which was in the format with which the engineers are well acquainted and a method based on procedures which were also established and well understood. The assessment route can be applied to ferritic steel structures of any discipline but its application has been orientated to pressure bearing components of Nuclear Class 1 status. The route deals with the treatment of failure in the small-scale and large-scale yielding regimes and takes account of fatigue crack growth. It does not deal with the arrest of a fast running crack, components which operate at temperatures at which time dependent creep effects are significant, stress corrosion cracking or impact loadings.

5.1. The Procedure

The procedure proposed by Harrison et al.[1] is

(1) determine actual flaw configuration from measured indication;
(2) resolve actual flaw into a simple shape;
(3) determine stresses at the location of flaw for all normal emergency and faulted conditions;
(4) calculate stress intensity factors for each condition;
(5) determine collapse load for structure containing flaw;
(6) determine necessary material properties;
(7) evaluate (a) ratio $K_I/K_{Ic} = K_r$ and (b) ratio of applied load/collapse load $= S_r$;
(8) Factorise if necessary to define position of assessment point K_{ra} on the failure assessment diagram;
(9) for acceptance the assessment point must lie within the failure assessment line;
(10) determine sensitivity of position of assessment point to flaw size;
(11) determine a_f—maximum size to which detected flaw is calculated to grow during life of component, a_c—minimum critical flaw size under operating conditions, a_i—minimum critical flaw size under emergency and faulty conditions; and
(12) determine acceptability of detected flaw.

The failure assessment route is one approach to assessing the integrity in structures containing defects. Out of the methods available a route has been selected for establishing the criticality of any flaw and examining the rate of growth of such a flaw. Varying degrees of sophistication can be used and it is hoped that the more simple methods prove to be satisfactory in the majority of cases. Harrison et al.[1] do not claim that the route is fully established. Through status notes they describe the extent to which each step is either fully proven scientifically, is established in theory only or merely represents the best judgement of those who have contributed to the formulation of this code of practice. The procedures are continually under review and are updated when knowledge and techniques are improved.

REFERENCES

1. HARRISON, R. P., MILNE, I. and LOOSEMORE, K., Assessment of the integrity of structures containing defects, 1977. CEGB Report R/H/R6-Rev 1.
2. DOWLING, A. R. and TOWNLEY, C. H. A. *Int. J. of Pressure Vessels and Piping*, 1975, **3**(3), 77.
3. BILBY, B. A., COTTRELL, A. H. and SWINDEN, K. H. The spread of plasticity from a notch, *Proc. Roy. Soc.*, 1963, **A272**, 304–14.
4. IRWIN, G. R., Relation of stresses near a crack to the crack extension force, *Proc. of 9th Int. Congress of Applied Mechanics*, Brussels, 1957, **8**, 245.
5. GRIFFITH, A. A., The phenomenon of rupture and flaws in solids, *Phil. Trans. Roy. Soc. London*, 1920, **A221**, 163.
6. HEALD, P. T., SPINK, G. M. and WORTHINGTON, P. J. Post-Yield Fracture Mechanics, *Mat. Sci. Eng.*, 1972, **10**, 129.
7. DARLASTON, B. J. L., HARRISON, R. P. and MILNE, I., Evaluation of flaw indications. *Proc. 3rd Conf. on Periodic Inspection of Pressure Vessels*, 1976, London, Inst. Mech. Engineers.
8. MERKLE, J. G. Pretest analysis information for HSST program intermediate test vessel V5, ORNL, 1975.
9. WITT, F. J. Fourth National Symposium on Fracture Mechanics, 1970, Carnegie-Mellon University.
10. RASHID, Y. R. and GILMAN, J. D. 2nd Conf. Structural Mechanics in Reactor Technology, 1971, Berlin.
11. ROOKE, D. P. and CARTWRIGHT, D. J. *Compendium of Stress Intensity Factors*, 1976, HMSO, London.

Chapter 5

RECENT DEVELOPMENTS IN METHODS FOR MATERIAL TOUGHNESS ASSESSMENT

C. G. CHIPPERFIELD

UKAEA, Risley Nuclear Power Development Laboratories, Risley, UK

SUMMARY

This chapter summarises and comments on current experimental practice and methods of analysis for the determination of the static and dynamic fracture toughness of structural materials. Particular emphasis is given to a discussion of several concepts which have been proposed for the assessment of material toughness in the presence of significant plasticity (including the J contour integral, crack opening displacement and equivalent energy approaches), their relative merits and their interrelationship. Further discussion is also given to the possibility of correlating toughness values with uniaxial tensile and Charpy V-notch impact data and to recent advances in resistance R-curve methodologies.

Despite encouraging progress in the R-curve field, these studies have not, as yet, reached a stage where these approaches can be reliably employed in assessing the significance of defects in structural components. It is recommended, therefore, that fracture toughness tests should currently focus on the conditions for first crack advance, irrespective of the method of analysis which is subsequently adopted; furthermore, for reasons of constraint, such tests should be performed on a specimen which is equivalent in thickness to the structural component of interest.

NOMENCLATURE

a	crack length
Δa	crack extension
B	test piece thickness

C	compliance
C_v	Charpy impact energy
D	semi-gauge length for a CCT specimen
E	Young's modulus ($E' = E/(1 - v^2)$)
e	strain
e_Y	yield strain
G	strain energy release rate
J	J contour integral
J_e, J_p	elastic and plastic components of J, respectively
K	stress intensity factor
K_A	apparent critical stress intensity factor
K_{Ic}	plane strain, linear elastic stress intensity factor
K_D	fast fracture toughness
K_d	dynamic critical stress intensity factor
K_{Id}	plane strain stress intensity factor under dynamic loading conditions
M	constraint factor
n	work-hardening exponent
P	load
P_{GY}	general yield load
R	crack growth resistance
S	loading span for a three point bend specimen
s	inclusion spacing
T	temperature
U	total absorbed energy (less extraneous components)
U_e, U_p	elastic and plastic components of U, respectively
V_{gt}, V_{ge}	total and elastic clip gauge displacements, respectively
W	test piece width
Y	compliance function
z	knife edge thickness
Δ	load point displacement
Δ_c, Δ_{nc}	components of Δ resulting from the presence of the crack and due to the uncracked body, respectively
δ	crack opening displacement (COD)
δ_c, δ_i	critical and initiation values of COD, respectively
σ_Y	static yield strength
σ_f	flow stress
ρ	notch root radius
ρ_0	limiting notch root radius
v	Poisson's ratio

1. INTRODUCTION

A key parameter in all of the preceding descriptions of methods for assessing the significance of flaws is that characterising the fracture toughness of the material. Established testing procedures are available for measuring reproducible values of plane strain fracture toughness, K_{Ic} provided that the region of plasticity associated with the stress concentrator is small in relation to the dimensions of the test piece.[1,2] These specimen size requirements, however, are such as to necessitate that all external dimensions exceed approximately fifty times the radius of the plane strain plastic zone, which at fracture is proportional to the ratio $(K_{Ic}/\sigma_Y)^2$ where σ_Y is the yield strength of the test material; for high toughness, low yield strength materials, therefore, the specimen size required to permit the measurement of valid K_{Ic} data becomes prohibitively large. Approximate extensions to this K concept have been advocated which essentially assume that the actual crack length has been increased by an amount equal to the radius of the plastic zone but the usefulness of this equivalent elastic crack concept, however, becomes questionable as general yield is approached because of a plasticity interference effect with the free surfaces of the specimen. Indeed, as discussed in earlier chapters, the accuracy of the equations for the plastic zone size can also be questioned since finite element analyses have shown that the size and shape of the plasticity region is dependent, at a particular applied K level, on the specimen geometry and loading configuration.[3]

Several toughness concepts have been described in the preceding chapters which are intended for use in the regime of more extensive plasticity and the evaluation of the validity of these analysis methods is the subject of ongoing research. Three of these concepts, namely the J-integral, equivalent energy and the crack opening displacement (COD) approaches, have received rather more experimental and theoretical interest and much of the following discussion is centred on these toughness assessment methods and their interrelationship. Although each of these has been discussed at length earlier, certain relevant features will be summarised again to make this chapter more comprehensible in itself. These techniques generally employ fatigue precracked specimens in order to simulate the worst conditions of notch acuity that could occur in a structural component and generally similar geometries and loading systems as for the K_{Ic} test. Fatigue precracking is required to be conducted in the range of applied stress where the K concept is operative and at a load very much less than that anticipated subsequently for fracture. Subsequent testing is conducted under conditions of temperature and strain rate relevant to the component of interest.

These toughness concepts are specifically aimed at characterising the conditions for first crack advance from the stress concentrator and thus rely on the accuracy with which the initiation point can be determined. A following section of this paper therefore examines methods whereby the onset of crack growth can be determined particularly for materials exhibiting slow stable crack extension for which the initiation point is generally not discernible from conventional test records. In addition, some discussion is centred on the significance, effect and current understanding of the stable crack growth process since an initiation toughness measurement may in some cases represent an over-conservative approach in respect of a material's resistance to unstable fracture.

2. J-INTEGRAL CONCEPT

A line integral of the form:

$$J = \int_\Gamma \left(F\,\mathrm{d}y - T\frac{\mathrm{d}u}{\mathrm{d}x}\mathrm{d}s \right) \tag{1}$$

has been advocated and applied to crack problems by Rice.[4] Γ is a contour taken counterclockwise from the lower to the upper surface of the notch (Fig. 1), F is the strain energy density, T is the tension vector perpendicular to the arc length $\mathrm{d}l$ and u is the displacement in the x direction. For a non-linear or linear elastic material it can be shown that the magnitude of J is independent of the choice of Γ for a straight crack along the x axis and in the absence of time dependent thermal stresses; in addition it can be shown that for a linear or non-linear elastic material, eqn. (1) is equivalent to:[5]

$$J = -\frac{1}{B}\frac{\mathrm{d}U}{\mathrm{d}a} \tag{2}$$

where U/B is the total potential energy of the system per unit thickness.

For a linear elastic material, therefore, the contour integral corresponds to the elastic strain energy release rate G. Recent investigations[6,7] have focused on the possibility of extending the J-integral concept to characterise elastic–plastic fractures. In such cases, it has been shown by finite element analyses that the J integral remains essentially path independent for materials obeying the laws of incremental plasticity and that the integral (eqn. (1)) and the energetic (eqn. (2)) definitions for J are still approximately equivalent, at least for the limited number of geometries that have been studied.[8] However, because of the irreversibility of plastic deformation, J strictly loses its energetic interpretation as the energy available for fracture although it can perhaps still be thought of as the difference in energy imparted to a given geometry containing incrementally

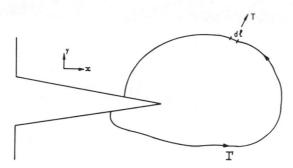

FIG. 1.   Crack tip co-ordinate orientation and arbitrary J-integral contour.

different crack lengths and may therefore still be regarded as a characterising parameter for crack tip damage. The difference in unloading characteristics between an elastic (linear of non-linear) material, where the loading and unloading curves will be identical, and that for an elastic–plastic material essentially implies that the J contour integral should, if it is applicable at all to elastic–plastic materials, be limited to situations with no unloading. Since crack extension leads to relaxation and unloading in the regions behind the growing crack tip, the requirement for no unloading implies that J should be restricted to monotonic loading situations and only be used to characterise events leading up to first crack extension.

Experimental determinations of J may either be made by monotonically loading specimens with differing crack lengths and employing eqn. (2) to estimate J[6,7] or by means of various estimation methods which have been proposed in the literature for commonly used specimen geometries.[9−12] Most of these estimation formulae require the measurement of load, P, and load-point displacement, Δ, during the toughness test and one of the simpler estimation procedures is that derived for an elastic–perfectly plastic material which relates J to the total elastic (U_e) and plastic (U_p) energies (Fig. 2) absorbed by the specimen. For the single-edge notched (SEN) bend, compact tension (CT) and centre-cracked tension (CCT) specimen, the relationship is of the form:[10]

$$J = \frac{1}{B(W-\gamma a)} (\eta_e U_e + \eta_p U_p) \tag{3}$$

where $\gamma = 1$ for bend and CT geometries, and 2 for CCT specimens, and where η_e and η_p assume the values given in Table 1 as a function of a/W and loading configuration. While the energy components U_e and U_p represent

TABLE 1

VALUES OF THE CONSTANTS η_e AND η_p FOR THE APPLICATION OF EQN. (3)[10]

Specimen type	η_e						η_p (all a/W)
	$a/W = 0.2$	0·3	0·4	0·5	0·6	0·7	
Three-point bend $S/W = 4.0$	1·4	1·7	1·9	2·0	2·0	1·9	2·0
CTS $H/W = 0.6^1$	3·7	2·7	2·4	2·3	2·2	2·2	2·0
CCT $D/W = 2.5$	0·20	0·28	0·34	0·38	0·40	—	1·0

the actual measured elastic and plastic energies less any extraneous energy components which may arise such as those associated with load point indentations and loading frame compliance effects, it is emphasised that both energy components otherwise refer to total absorbed energies. When a fracture toughness specimen deforms, the total load-point deflection, Δ, which is measured during the test, is composed of two components: firstly, that due to the presence of the crack, Δ_c, and that which would have arisen in the absence of the crack Δ_{nc} due to the compliance of the unnotched specimen. Both the displacement components can be accurately calculated and have been used to derive the η_e values in Table 1 as follows: because of the equivalence of G and J in the linear elastic regime, the values of η_e have been established using the relationship:

$$G = -\frac{P^2}{2B} \frac{dC_c}{da} = \frac{\eta_e U_e}{B(W-\gamma a)} \tag{4}$$

where C_c and C_{nc} are the compliances of the cracked and uncracked test piece and U_e is given by the area under the load, P, versus $(\Delta_c + \Delta_{nc})$ curve:

$$U_e = \frac{P^2}{2} (C_{nc} + C_c) \tag{5}$$

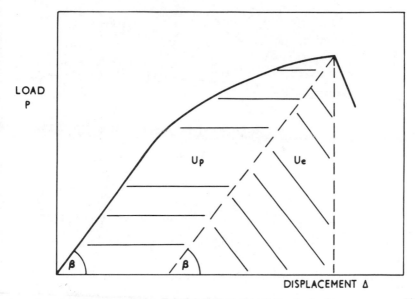

FIG. 2.   Construction for separating U into elastic and plastic components.[10]

It is clear that if the load versus Δ_c curve is required to be analysed by the area procedure of the form of eqn. (3), then if U_e is briefly redefined as the elastic energy under the load/Δ_c curve, $(0 \cdot 5\, P^2 C_c)$, the η_e values of Table 1 will be no longer relevant and will, because C_{nc} is always greater than zero, in general assume larger values. On the other hand, η_p will remain unchanged. As an example, for a three-point bend specimen with a loading span, S, to test piece width, W, ratio of $4 \cdot 0$, the value of η_e required in eqn. (3) for the analysis of a load/Δ_c curve would be $5 \cdot 05$ and $3 \cdot 02$ for an a/W ratio of $0 \cdot 3$ and $0 \cdot 5$, respectively. The differences between these values and those of Table 1 illustrate the increasing proportion of the total deflection Δ arising from that of the uncracked beam as the a/W ratio decreases.

It is to be noted that because the estimation methods for J are usually based on its energetic interpretation (eqn. (2)), the requirements for monitoring test piece displacement differ from that advocated for K_{Ic} or COD fracture toughness tests. An alternative estimation procedure for J has recently been advocated for three-point bend specimens ($0 \cdot 25 < a/W < 0 \cdot 60$), however, which utilises clip gauge displacement monitoring of the crack mouth opening.[10] As in the case of the energy based method (eqn. (3)), the value of J at the failure load, P_f, is calculated by separating the toughness parameter into elastic and plastic components; J_e is calculated

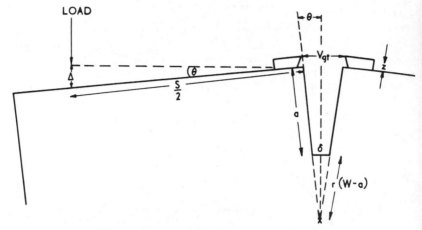

FIG. 3.   Relationship between V_{gt}, Δ and δ.

on the basis of standard compliance expressions and the load, P_f, as follows:

$$K_f = \frac{P_f Y}{BW^{1/2}} = (E' J_e)^{1/2} \tag{6}$$

and J_p is calculated by assuming no work-hardening beyond the general yield load, P_{GY} and by relating Δ to clip gauge displacement by assuming rigid body rotation about a point a distance $r(W-a)$ ahead of the crack tip (Fig. 3):

$$J_p = \frac{2P_{GY}}{B(W-a)} \left(\frac{W}{a + r(W-a) + z} \right) (V_{gt} - V_{ge}) \tag{7}$$

where V_{gt} and V_{ge} are the total and elastic clip gauge displacements as measured a distance z above the specimen. The recommended values for r are specified as $0\cdot45\,(a/W \leq 0\cdot45)$ and $0\cdot4\,(a/W > 0\cdot45)$.

A recent summary and comparison of the various estimation procedures[13] has tentatively endorsed, on the basis of all known data, the use of both the above methods for specific specimen configurations and ranges of a/W. For the purposes of the comparison it was assumed that the compliance method for deriving J values by monotonically loading specimens containing successively longer cracks (eqn. (2)) gave accurate values for J and these were then compared with estimated values. Equation (3) was tentatively recommended for use on CT specimens employing an

a/W ratio in the range 0.5–0.7 and for SEN bend specimens with an a/W ratio in the range 0.3–0.5. On the other hand, the use of eqns. (6) and (7) was recommended throughout its range of validity on SEN bend specimens since, by assuming the absence of work-hardening beyond general yield, the method generally appears to provide conservative values for J.

An experimental procedure for J testing has recently been proposed by Begley and Landes,[14] for which the use of deeply notched ($a/W \geq 0.6$) CT or SEN bend specimens is advocated. For these test pieces, eqn. (3) reduces approximately to:

$$J = \frac{2U}{B(W-a)} \tag{8}$$

and eqn. (8) is used to characterise the conditions for crack initiation. If initiation and instability are coincident, the critical value of J, usually termed J_c, is determined from the total energy, U, absorbed at instability (less extraneous energy components). If, however, instability is preceded by slow stable crack growth it is necessary to test a series of samples to progressive smaller deflections before unloading. The specimens are then heat tinted and broken open to enable the extent of slow crack growth to be measured. A resistance R-curve is then constructed of J versus crack extension, Δa (Fig. 4). The measured values of Δa, however, generally include both true stable crack extension and a component which arises from progressive crack tip blunting. The latter region is generally termed a 'stretch zone' and Begley and Landes argue that the extension of the original defect by this blunting mechanism obeys the relationship:

$$J = 2\sigma_f \Delta a \tag{9}$$

where σ_f is usually taken to correspond to the arithmetic mean of the yield and ultimate tensile strength of the test material. The crack tip blunting line is thus superimposed on the resistance curve (Fig. 4) and the initiation value of J, J_i, is then defined as the value at the intersection of the blunting line and the experimental resistance R-curve.

With regard to the choice of specimen size, it has been suggested that the J_c or J_i value obtained by the above procedure will be representative of the conditions for crack initiation in larger specimens provided that the following size requirements are satisfied:

$$a, B, W-a > \frac{\alpha J_i}{\sigma_f} \tag{10}$$

where α lies in the range 25–50.[14] While this relationship appears to be

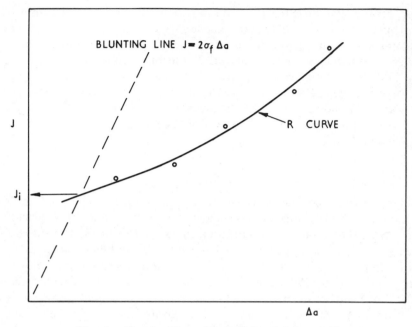

FIG. 4.   Construction advocated for defining J_i.[14]

consistent with some reported data on materials which exhibited a ductile fracture mode,[14,15] for which the size independency of J_i is reasonably well established,[14-16] it is not at all certain whether such geometry requirements are still adequate for brittle failure modes (e.g. transgranular cleavage). Indeed, some data suggest, in line with reported COD data,[17] that small specimen data can provide optimistic estimates of the failure conditions for larger specimens.[14,18]

An explanation for this behaviour has been proposed[18] in which it is argued that since larger specimens sample more material they are more likely to contain locally embrittled areas than smaller specimens. This explanation, however, has yet to be substantiated by the testing of a statistically significant number of small specimens. A further point of caution concerns the advocated blunting equation, the derivation of which assumes plane stress deformation and that crack tip blunting proceeds along 45° slip lines at the crack tip. It is by no means certain that this equation will be universally relevant to all specimen configurations and to all materials. Indeed, the angle of slip line inclination has been predicted[19] and observed[16,20] to differ from 45°. It is to be noted with this in mind that small errors in the slope of the blunting line in Fig. 4 can give rise to significant errors in J_i particularly for materials which exhibit a steep R-curve.

As far as is known to the author, J_c or J_i values have not been widely employed for specifying critical defect sizes for structural components. While this could be achieved by conducting finite element analysis on the component design, this course of action seems unlikely to be followed extensively due to cost considerations. A recent paper[21] has suggested that, for contained yielding situations, a linear elastic approach is all that is required as long as the defect size established is reduced by an amount equal to the plastic zone size. For this procedure, J_c is measured for a test piece of equivalent thickness to that of the structure and is converted to a notional K_c value via the relation:

$$K_c^2 = E'J_c \tag{11}$$

An alternative design procedure[22] has been advocated in the USA and which closely resembles the design curve advocated for application of the COD concept:[23]

$$J = e^2 E\pi a \qquad (e \leq e_Y) \tag{12a}$$

$$J = E\pi e_Y(2e - e_Y)a \qquad (e > e_Y) \tag{12b}$$

It is emphasised, however, that both assessment methods have not, as yet, been widely used for the specification of critical defect sizes. Indeed, the question of whether either procedure can still be employed for defect evaluation in the presence of thermal stresses, which essentially invalidate the J approach, is as yet uncertain.

3. EQUIVALENT ENERGY METHOD

The equivalent energy concept proposed by Witt[24,25] represents a simple empirical approach for relating the fracture conditions in geometrically similar models. When applied to toughness specimens,[26] a notional stress intensity factor K_{IcB} is determined at the point of failure irrespective of the extent of plasticity. B refers to the specimen thickness and the subscript I refers to mode I opening of the crack but does not necessarily imply plane strain deformation. K_{IcB} is established by defining a fictitious failure load P^* and displacement Δ^* at the point along the extended, elastic load/displacement portion of the test record (Fig. 5) at which the area under

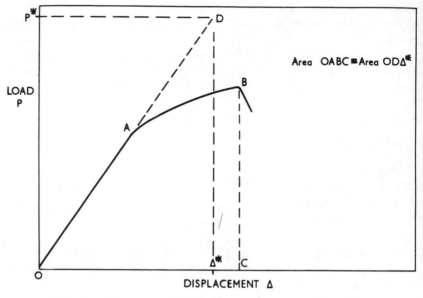

FIG. 5.   A summary of the equivalent energy procedure.[26]

the experimental load/displacement curve up to maximum load is
equivalent to the energy $0.5\,P^*\Delta^*$. K_{IcB} is then evaluated for a CT or bend
specimen via the relationship:

$$K_{IcB} = \frac{P^* Y}{B\sqrt{W}} \tag{13}$$

where Y is the relevant compliance function. It has been argued in several
instances that values of K_{IcB} so determined have represented lower bound
estimates of the valid linear elastic toughness K_{Ic} value and which tend to
approach K_{Ic} as the thickness B is increased.[27,28]

   It is to be emphasised that in the original formulation of the equivalent
energy procedure the experimentally determined failure point is taken as
that corresponding to maximum load and the question of whether stable
crack growth has occurred prior to this load is ignored. For the purpose of
the following brief discussion, however, a brittle failure mode is considered
where the conditions for initiation and test piece instability coincide.
Considering the fact that the J-integral and equivalent energy concepts are
both broadly based on energy considerations, it is worthwhile to examine
the relationship between the two procedures. Little guidance is specified

with regard to the choice of load-point displacement that is required in the equivalent energy method:[29] that is, whether in fact the displacement to be considered is that due solely to the presence of the crack, Δ_c, or whether the total displacement, less extraneous components, Δ, is the important parameter. Consider, for example, a load versus load-point displacement curve for an elastic–plastic material with no work-hardening as an approximation to that for a slowly work-hardening material. If crack initiation and instability occur simultaneously so that the measurement points for establishing toughness on the basis of the J-integral and equivalent energy concepts are identical then the value of J_c at a point beyond general yield using eqn. (3) would be:

$$J_c = \frac{\eta_e U_e + \eta_p U_p}{B(W-a)} \qquad (14)$$

Because of the compatability of eqn. (3) with K in the elastic regime ($U_p = 0$), the value of G_{IcB} where

$$K_{IcB}^2 = E' G_{IcB} \qquad (15)$$

which would be calculated on the basis of the equivalent energy concept can conveniently be derived directly from eqn. (3) as:

$$G_{IcB} = \frac{\eta_e(U_e + U_p)}{B(W-a)} \qquad (16)$$

since the total energy ($U_e + U_p$) is now assumed to be all elastic energy (Fig. 5). Clearly, eqns. (14) and (16) are only consistent and hence the J and equivalent energy procedures are only in exact agreement (i.e. $J_c = G_{IcB}$) when the values of η_e and η_p are equivalent for the test piece geometry under consideration. It is interesting to note that the test pieces which have generally been employed for the equivalent energy procedures have been deeply notched CT specimens for which the identity $\eta_e = \eta_p$ is approximately obeyed (Table 1). This may partially explain why the equivalent energy has been retained as an alternative method of analysis. In general, however, it is to be noted that the J and equivalent energy approaches are not consistent (Table 1). If $\eta_e > \eta_p$ then the equivalent energy approach will provide K values in excess of those implied by the J method, whereas if $\eta_p > \eta_e$ the converse is true.

Returning to the discussion in the previous section concerning the effect of the choice of load-point displacement measurement (i.e. Δ or Δ_c) on values of η_e, it is apparent that this choice can affect the agreement or lack of agreement between the J and equivalent energy procedures. For a shallow notched SEN bend specimen ($a/W = 0·3$) the value of η_e required in eqn. (3) is either 1·7 (Table 1) if the load/Δ curve is adopted or 5·05 if the load/Δ_c trace is considered, while η_p retains the value 2·0 for both approaches. It is clear from the previous discussion, therefore, that for such a specimen an equivalent energy analysis will only provide a value of K_{IcB} below that implied by the J procedure if the load versus *total* load-point displacement Δ (i.e. $\Delta_c + \Delta_{nc}$) curve is employed.

The preceding discussion has considered wholly brittle conditions for which stable crack growth was assumed to be absent. For such conditions it has been noted that the J and equivalent energy procedures are consistent in a restrictive range of|specimen geometries and hence it is to be expected that similar effects of test piece thickness and constraint should be experienced for such geometries and for both analyses. The fact that, for such brittle failure modes, small specimen tests have led to J_c values in excess of those for larger specimens appears to cast considerable doubt on the statement that K_{IcB} values represent lower bound estimates|of K_{Ic} in the brittle regime. In addition, in situations where slow crack growth occurs with increasing load as is generally observed in small specimens of high toughness material, pseudo-J values at maximum load instability, J_m, evaluated on the basis of energy up to maximum load can exceed values of J_i by a considerable amount. Since equivalent energy procedures advocate the use of the energy up to maximum load regardless of whether slow crack growth has occurred, it would seem that, by acknowledging the similarity between K_{IcB} and J_m measurements so derived, similar doubts concerning the lower bound nature of such K_{IcB} measurements are also of relevance for this temperature regime.

The logical extension to these arguments appears to be that values of K_{IcB} measurements should not be generally used for assessing the significance of defects in structures. If the lower bound nature of such results is subsequently substantiated, for example by suitable choice of specimen geometry and a requirement for initiation detection, defect assessments could perhaps be accomplished by means of such K_{IcB} measurements, through the use of standard elastic analysis routes and incorporating a plastic zone correction factor on defect size.

4. CRACK OPENING DISPLACEMENT (COD) METHOD

Following the proposals made independently by Wells[30] and Cottrell[31] that the critical crack tip opening displacement, δ, may adequately characterise the conditions for fracture under constant conditions of constraint but irrespective of the extent of yielding, a considerable amount of effort has gone into the development of a test technique for assessing material toughness based on the COD concept. The recommended test procedure is documented in a current British Standard Draft for Development[32] and essentially requires the measurement of load and crack mouth opening during the toughness test. The latter measurement is most conveniently made by the mounting of a double cantilever beam clip gauge across the crack mouth and located on knife edges of thickness z (Fig. 3). Recommended test piece geometries are advocated and the problems associated with the effect and choice of test piece thickness, as summarised in the previous sections, are largely circumvented by the specification that the thickness of the specimen is to be equivalent to that of structural component of interest, thereby ensuring a reasonable simulation of the in-service constraint.

The clip gauge displacement, V_{gt}, at the onset of crack extension is required and may be interpreted in terms of crack tip opening for post-yield conditions by assuming rigid body rotation about a point a distance $r(W-a)$ ahead of the crack tip. A recommended value for r advocated in the British Standard document is $\frac{1}{3}$ and the relationship between V_{gt} and COD, δ, can be shown to be given by simple geometry as:

$$\delta_c \text{ or } \delta_i = \frac{(W-a)}{(W + 2a + 3z)} V_{gt} \qquad (17)$$

where δ_c and δ_i represent the usual nomenclature for critical COD and the COD at the initiation of slow, stable crack growth, respectively. The basic assumption in the derivation of eqn. (17) is that the apparent centre of rotation is located a constant distance ahead of the crack tip throughout the test, whereas the value of r at low loads will be close to zero. Clearly, therefore, there are limitations on the use of eqn. (17) in the region of general yield and these are outlined in the Draft recommendations together with alternative formulae for the derivation of COD.

The general philosophy embodied in the COD testing specification is that by utilising a specimen of a thickness equivalent to that of the structural component of interest, the constraint existing in the specimen will be closely representative of that in the structure. COD results particularly for ferritic materials in the temperature transition range have supported this general philosophy[17] by indicating that at the same temperature, lower δ_c values are exhibited by thicker test pieces. On the upper shelf regime however where the fracture mode is wholly ductile microvoid coalescence, the value of

COD at ductile crack initiation, δ_i, appears to be relatively independent of test piece size[16,33,34] above some critical value B_{min} which, it has been suggested, is approximately given by:[35]

$$B_{min} \simeq 25\, \delta_i \qquad (18)$$

A possible reason for this behaviour has been proposed and relates to the fact that the notch tip fracture strain, of which the COD is a measure,[33] appears to be relatively insensitive to the magnitude of the constraint.[36] This does not necessarily mean, however, that a δ_i value established for a small sample will necessarily characterise the conditions for fracture initiation in a larger sample, since the fracture modes exhibited by the two samples may differ.

Examination of the theoretical relationship between the J contour integral and COD by means of finite element analysis has been hampered largely by the fact that for certain specimen geometries, and particularly those subject predominantly to tension, the COD is difficult to define; the blunting stress concentrator assumes a more rounded profile at the notch tip when compared with that found for bend geometries.[37] However, taking the simplest case of a three-point bend specimen of rigid-plastic material for which the general yield load, P_{GY}, is given by:[38]

$$P_{GY} = 1 \cdot 5\, \sigma_Y \frac{B}{S} (W{-}a)^2 \qquad (19)$$

where S is the loading span, J at a particular displacement, Δ, is then given from eqn. (2) by:

$$J = -\frac{1}{B}\frac{dU}{da} = -\frac{\Delta}{B}\frac{dP_{GY}}{da} = 3\,\sigma_Y(W{-}a)\frac{\Delta}{S} \qquad (20)$$

since U is equivalent to $P_{GY}\Delta$. Also, by assuming rigid body rotation about a point $(W{-}a)/3$ ahead of the crack tip, δ and Δ are related by (Fig. 3):

$$\frac{3\,\delta}{2(W{-}a)} = \frac{2\,\Delta}{S} \qquad (21)$$

and therefore eqn. (20) reduces to:

$$J = 2 \cdot 25\, \sigma_Y \delta \qquad (22)$$

Finite element analyses have essentially confirmed the value of the constraint factor $M \sim 2$ for bend geometries but have indicated that the relationship can alter depending on the loading configuration; for example, a value of M equal to $1 \cdot 16$ is predicted for the centre-cracked tension geometry and such results therefore suggest that the J and COD are mutually exclusive concepts.[39] Such predictions have not been sub-stantiated by a limited series of experiments on CCT and SEN bend specimens where the values of J_i and δ_i were the same in both tension and bending.[40] The exact value of M has however been shown to vary from about unity to 2·6 as a function of the work-hardening capacity of the test material and the extent of yielding of the specimen.[16,40] This experimentally observed variation in M could at least in part be due to the uncertainties surrounding the determination of COD values in the neighbourhood of general yield; on the other hand, direct COD measurements using rubber infiltration moulding techniques have confirmed this variation in M from between unity and in excess of 2 as a function of work-hardening capacity.[40] At the present time, therefore, the evidence suggests that an initiation toughness value (be this J_i or δ_i) determined in a bend geometry is equally applicable to a specimen or indeed a structural component subject to tension.

The general relationship between J and δ of the form:

$$J = M\sigma_Y\delta \tag{23}$$

implies that the effects of test piece geometry and size should be broadly similar for both toughness parameters. It is interesting to note, therefore, that eqns. (10) and (18) are equivalent for $\alpha = 25$ in terms of specimen thickness requirement for an assumed value for M of unity.

However, as a result of the observed effect of test piece thickness on δ_c in the temperature transition range and uncertainties on reaching the upper shelf toughness concerning whether a large and small specimen will exhibit the same ductile fracture mode, it would seem to be obligatory at the present time to test a specimen equivalent in thickness to the structural component of interest, irrespective of whether the toughness is evaluated in terms of J or COD. As has been previously noted this approach is specified in the Draft recommendations for COD testing and the critical or initiation value of crack opening displacement established by these procedures may be used directly to establish the significance of defects by means of a semi-empirical design curve.[23]

5. THE DETECTION OF INITIATION

The requirement for the ability to assess the resistance of a material to the initiation of fracture, even when in a specimen test subsequent crack extension is inherently stable, results from the fact that a detailed understanding of the instability process has yet to be obtained (see Section 8). The current procedure generally employed to establish the point of first crack extension is that outlined in Section 2 and generally requires the preparation and testing of perhaps six similar samples in order to obtain a resistance curve and hence, by extrapolation to zero crack extension, a single value of δ_i or J_i as appropriate. Because of the cost involved in such a procedure, there is an obvious need for a reliable technique which permits the detection of initiation from a single specimen test.

Crack monitoring techniques have been used extensively in fatigue studies but less widely to detect initial crack extension in fracture toughness tests. However, the methods available for such crack detection include:

(i)   the acoustic emission technique, which attempts to detect stress waves emitted from the crack tip region as the defect extends,

(ii) the compliance procedure, which attempts to detect crack extension by identifying a change in the test piece stiffness as the defect extends, and

(iii) the electrical potential method for which a constant current is applied to the specimen and the potential either side of the crack is monitored during the test. Either a direct (dc) or alternating (ac) constant current source may be employed: for the dc method, crack extension gives rise to an increasing current density in the ligament $(W-a)$ and is reflected in an alteration in the potential output across the crack faces; the ac technique on the other hand utilises the skin effect by employing a current of a suitably high frequency and crack extension gives rise to an increasing current path around the defect which is reflected as a change in potential across the crack faces. The dc method is generally used only for smaller specimen sizes because of the high current density required in the ligament.[32] The ac method is amenable, however, to all specimen sizes since the current skin depth is dependent only on the frequency of current input and the resistivity and magnetic permeability of the test material and is independent of test piece dimensions.

A recent investigation has focused on the ability of such techniques to detect the onset of ductile crack extension in generally yielded fracture toughness specimens.[16] For such specimens, a high strain field is developed in front of the crack tip over a distance of approximately twice the applied COD; this effect gives rise to an increase in resistivity of the crack tip region over and above that caused by slow crack growth. Detection of slow crack growth by means of the electrical potential method is thus complicated by the appearance of this intensely strained region, although for lower toughness (δ_i) materials the smaller extent of this region increases the ability to detect the initiation of slow crack growth. The sensitivity of the compliance and electrical potential detection methods is additionally dependent, for given experimentation, on the rate of crack advance once initiation has occurred. If one considers that a particular detection apparatus can discern a change in crack length of Δa and that the applied COD, δ, increases linearly with crack extension (i.e. $d\delta/da = $ constant) then the detection method will overestimate the initiation COD by an amount $\Delta a(d\delta/da)$.

Detection of the actual initiation event is thus more nearly achieved when the slope of the resistance curve ($d\delta/da$) is small, as is generally the case for lower toughness materials. A compliance procedure has also been investigated by Clarke et al.,[41] for which the compliance alterations were monitored by interrupting a static test at frequent intervals, unloading the test piece by a maximum of 10% of the maximum load and monitoring the slope of the unloading load/Δ curve. The technique appeared to produce satisfactory results although the conclusions are not necessarily at variance with above conclusions since the material tested was of high yield strength and thus presumably exhibited a slowly rising R-curve. The possibility of an effect of such an unloading procedure on J measurements should not, however, be discounted: a more attractive method which obviates the need for this unloading procedure would appear to be to monitor two separate displacement measurements (e.g. V_{gt} and Δ) on the specimen and essentially measure the position of the apparent centre of rotation of the specimen during the test since, as the crack grows, the centre of rotation appears to move in the same direction. The applicability of this latter technique is presumably limited, however, to situations in which crack initiation occurs beyond general yield where, in the absence of crack extension, the position of the rotational centre is constant.

With regard to acoustic emission monitoring, broadly similar conclusions are relevant with initiation detection being more easily achieved for materials which exhibit rapid or sudden crack extension be this on a macro- or microscale. In the ductile upper shelf regime, fracture occurs by the formation, growth and coalescence of microvoids and the ability to detect the initiation event rests principally on whether or not the final coalescence event is an inherently 'noisy' process. In low strength, high toughness steels it would seem that the coalescence process results from gradual plastic flow localisation between the crack tip and a favourably oriented void and gives rise to similar emissions to those emanating from the surrounding uniformly strained region; the detection of initiation in such cases by acoustic techniques is therefore extremely difficult. In some higher strength steels in the ductile regime, however, the detection of initiation is somewhat easier[16] since crack growth tends to proceed by a locally unstable mechanism involving the rapid 'unzipping' of microvoids.[42]

In general, therefore, the detection methods summarised above are only able to detect accurately the onset of stable crack growth in specific instances and particularly when the rate of crack extension is large. The compliance and electrical potential methods, however, are still of use as a general experimental facility as they generally provide an estimate (albeit an unconservative one) of initiation from a single test record and may therefore reduce the number of samples that are required to develop a resistance curve of the type advocated for the J technique and which is subsequently extrapolated to zero crack extension.

6. PREDICTION OF TOUGHNESS VALUES AND TRENDS

As a result of the fact that direct experimental measurements of K_{Ic}, J_c and δ_c are expensive to conduct and therefore not generally used for quality control purposes, a considerable amount of effort has been expended over the years in attempts to correlate the results of simpler mechanical tests with material toughness and particularly K_{Ic}. Several recent advances and observations have been made in this area and some of these are summarised below.

6.1. Low Temperature Fracture

A number of correlations have been proposed for the relationship between the energy absorbed in a Charpy V-notch test, C_v, and the valid plane strain fracture toughness K_{Ic} in the low temperature, essentially brittle fracture regime.[43–45] The Charpy test piece is still used extensively for quality control and surveillance purposes and the ability to predict static fracture toughness values from a simple test would clearly be advantageous. The standard Charpy V specimen, however, employs a root radiused notch (tip radius equal to 0·25 mm), is only of 10 mm square section and is tested at high strain rates and the problems associated with the theoretical proof of the uniqueness of such relationships between C_v and K_{Ic} would appear to be immense. The effect of notch acuity on the apparent dynamic toughness, K_A, associated with a blunt notch has been investigated both experimentally[46] and theoretically[46,47] and has been shown to follow an approximately linear relationship with $\rho^{1/2}$ at a given temperature, where ρ is the notch root radius, down to a critical root radius, ρ_0, comparable with the grain size, below which $K_A = K_d$, where K_d refers to the value of toughness evaluated for a fatigue precracked specimen. Thus:

$$\frac{K_A}{K_d} = \left(\frac{\rho}{\rho_0}\right)^{1/2} \qquad (\rho \geq \rho_0) \qquad (24)$$

The effect of the differences in section thickness and strain rate in Charpy and K_{Ic} specimens is impossible to assess at the present time but presumably may be thought of as resulting in the translation of the toughness versus temperature curve along the temperature axis. A recent correlation method has been proposed by Marandet and Sanz[45] for the interpretation of K_{Ic} values from Charpy impact data. The approach is limited to ferritic steels for values of C_v up to 50 joules and has been substantiated by a study of a considerable number of different steels and heat treatment conditions: the advocated procedure is summarised as follows:

(i) the impact (C_v) transition curve is established
(ii) the general shape of the K_{Ic} transition curve is established from the relationship:

$$K_{Ic} = 19 \, (C_v)^{1/2} \qquad (25)$$

where C_v and K_{Ic} have units of joules and $MNm^{-3/2}$ respectively.

(iii) the temperature T_{100} at which K_{Ic} corresponds to $100 \, MNm^{-3/2}$ is given by:

$$T_{100} = 9 + 1 \cdot 37 \, (T_{28}^c) \, ^\circ C \qquad (26)$$

where T_{28}^c is the temperature at which a C_v value of 28 J is recorded.

(iv) the exact position of the K_{Ic} versus temperature curve developed in (ii) above is determined by the translation of the curve along the temperature axis until it passes through the point with coordinates

$$T = T_{100}, \quad K_{Ic} = 100 \, M \, Nm^{-3/2}$$

In broad terms, therefore, the approach is similar to that discussed earlier. Values of C_v are essentially used to develop the absolute shape of the K_{Ic} temperature transition curve and the curve is then translated along the temperature axis by an amount $T_{100} - T_{28}^c$ which can be positive or negative. Presumably the sign of the temperature shift required is governed by the relative magnitudes of two components, ΔT_{th} and ΔT_{sr}: firstly there is a translation to higher temperatures arising from the effect of thickness, ΔT_{th} (essentially converting K_{d} to plane strain K_{Id} values), and an accompanying translation to lower temperatures by an amount ΔT_{sr} to allow for strain rate differences (i.e. a conversion of K_{Id} to K_{Ic}).

It is instructive to attempt to develop an approximate relationship between C_{v} and K_{d} in the light of this method by assuming that the Charpy specimen fails in a brittle manner before net section yield and that C_{v} measurements reflect the initiation energy only. The apparent critical stress intensity factor, K_{A}, for the Charpy specimen is thus given by eqns. (3) and (6) as:

$$K_{\text{A}} = \left\{ \frac{1\cdot4\,EC_{\text{v}}}{B(W-a)} \right\}^{1/2} \tag{27}$$

By invoking eqn. (24), the equivalent K_{d} value for a fatigue precracked specimen of Charpy geometry may be estimated as:

$$K_{\text{d}} = \left\{ \frac{1\cdot4EC_{\text{v}}\rho_0}{B\rho(W-a)} \right\}^{1/2} \tag{28}$$

With $E = 175 \times 10^3\,\text{MNm}^{-2}$, $\rho = 0\cdot25\,\text{mm}$ and assuming that ρ_0 is equivalent to a typical grain size ($25\,\mu\text{m}$), evaluation of eqn. (28) for the Charpy geometry yields:

$$K_{\text{d}} = 17\cdot4\,(C_{\text{v}})^{1/2} \text{ (units as for eqn. (25))} \tag{29}$$

which is in reasonable agreement with eqn. (25) considering the assumption of a particular value of grain size. Equation (29) and the dependency of K_{d} on grain size does imply, however, that the above correlation method of Marandet and Sanz may not adequately describe the behaviour of a wide range of structural materials; for example, weld metals which were not included in the original work on which the method was based. It is to be noted, however, that K_{d} in eqn. (28) is not very sensitive to small variations in ρ_0.

6.2. Upper Shelf Regime

The fracture mode at higher temperature tends to be associated with widespread plastic deformation and the formation, growth and coalescence of holes emanating from second phase particles in the matrix. Unlike the initiation of brittle fracture, the total energy absorbed in fracturing a Charpy specimen, C_{v}, no longer approximates to the energy required to initiate fracture but also contains a significant portion which is associated with the propagation stage. Despite this aspect, however, a correlation between C_v and K_{Ic} values has been proposed although some of the toughness data quoted in support of the empirical equation were invalid according to current K_{Ic} testing specifications. The relationship advocated by Rolfe and Novak is:[48]

$$\left(\frac{K_{Ic}}{\sigma_Y}\right)^2 = 0.65\left[\frac{C_v}{\sigma_Y} - 0.009\right] \tag{30}$$

where K_{Ic}, σ_Y and C_v are in units of $MNm^{-3/2}$, MNm^{-2} and joules respectively and are each measured at $26\,°C$. By assuming a yield strain ε_Y equal to 1.3×10^{-3} and relating δ_i to J_c via a constraint factor M (eqn. (23)) and thence to K_{Ic} (eqn. (6)), eqn. (30) is expressible as:

$$\delta_i = \frac{0.87}{M}\left[\frac{C_v}{\sigma_Y} - 0.009\right] \tag{31}$$

where δ is in millimetres. In order to investigate the applicability of eqn. (31) to the ductile fracture regime, all known data are shown in Fig. 6 for a variety of ferritic and austenitic steels on the upper shelf (although not necessarily at $26\,°C$). It is apparent that most of data fall between the lines evaluated on the basis of eqn. (31) with $M = 1$ and $M = 2.5$, which approximately correspond to the extremes of the range anticipated for the constraint factor (Section 4). On a closer inspection, it appears that the materials which contained a relatively high volume fraction of second phase particles (i.e. an Enla high sulphur mild steel and a 316-type weldment) obey the $M = 2.5$ relationship while the comparatively cleaner wrought structural steels tend to fall nearer the $M = 1$ line. A possible explanation for this behaviour other than being an effect of constraint may lie in the fact that the resistance of a material to the initiation of ductile fracture as a function of notch acuity is dependent on the spacing of potential void nuclei in the material.[42] The initiation COD for a precracked test piece, δ_i, and that for a notch of radius ρ, $\delta_{i\rho}$, have been observed to be related, subject to the test piece dimensional requirements of eqn. (18) by a critical crack tip fracture strain, ε_f, as follows:

$$\frac{\delta_i}{\delta_{i\rho}} = \frac{s\varepsilon_f}{\rho\varepsilon_f} = \frac{s}{\rho} \tag{32}$$

For a given value of δ_i, constraint factor M and root radius ρ (equal to $0.25\,mm$), therefore, larger values of $\delta_{i\rho}$ and hence larger initiation energy

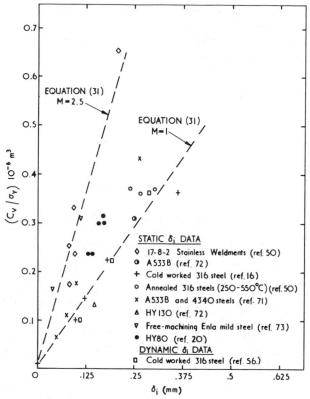

FIG. 6.   Examination of a possible correlation between δ_i and C_v.

components of C_v will be exhibited by materials with a smaller inclusion spacing, s. This argument therefore suggests that a unique relationship of the form of eqn. (31) is unlikely to characterise all materials although a relationship modified on the basis of inclusion content might be possible.

The data of Fig. 6, however, do indicate that an approximate correlation between C_v values and initiation general yielding toughness parameters may be possible on the upper shelf regime. Complications arise, however, because of the differences in notch acuity and strain rate as with empirical relationships established for the brittle fracture regime, although at upper shelf temperatures the yield strength of ferritic and austenitic steels is not expected to be very dependent on strain rate. An additional problem to such a correlation lies in the fact that on the upper shelf, the shear lip (plane stress) fracture energy becomes significant for a Charpy test piece whereas the δ_i data all refer to the onset of crack extension at specimen mid-thickness for test pieces satisfying the less restrictive dimensional requirements advocated for plane strain (eqn. (18)).

A relationship of the form of eqn. (31) subject to its substantiation by further data does provide, however, some insight into the anticipated effect of, for example, low temperature neutron irradiation, which is generally observed to 'give rise to matrix hardening (increasing σ_Y) and a reduction in upper shelf C_v values. Equation (31) would thus imply that neutron irradiation would lead to a decrease in δ_i values although in terms of J_i the effect might be at least partially offset by the concomitant rise in yield strength. However, little guidance can be obtained with respect to the effect of temperature on δ_i values since both C_v and σ_Y tend to reduce as the test temperature is increased. Recent indications are, however, that increasing test temperature can give rise to a moderate reduction in δ_i in several structural steels and consequently, because of the relationship between J_i and $\sigma_Y \delta_i$ and the concomitant effect of temperature of σ_Y, to larger percentage reductions in values of J_i.[49,50] Although such upper shelf toughness trends with temperature have not been observed in all the structural steels which have been studied to date,[18] these results do indicate that it should not be generally assumed that the initiation toughness observed on the upper shelf at, for example, 50 °C is equally applicable or represents a lower bound with respect to that exhibited at higher temperatures. The reason for the reductions in δ_i observed in several steels[49,50] with increasing temperature is not currently understood but may be related to the work-hardening rate of the material. Observations of the spread of yielding ahead of a stress concentrator in silicon–iron[51] have suggested that the strains developed at the crack tip are distributed over a width proportional to n^2, where n is the work-hardening exponent for material displaying a uniaxial true stress–true strain curve of the form:

$$\sigma = A\varepsilon^n \qquad (33)$$

where A is a material constant. Consequent predictions are that δ_i should vary in proportion to n^2 and such a relationship has been substantiated for the ductile fracture of aluminium alloys[51] and cold-worked stainless steels.[16]

It would appear, therefore, that some insight into the magnitude and trends of static toughness values can be gleaned from comparatively simple mechanical tests. The main problem with Charpy specimens, in particular, is that they provide a total fracture energy from a radiused notch for a test piece which does not, in general, adequately simulate the constraint and loading rate of interest. A correlation between C_v values and dynamic toughness values would reduce some of the uncertainty involved in these correlations and one such empirical equation has been developed by Barsom and Rolfe on the basis of dynamic Charpy toughness data evaluated for the lower temperature regime.[43] In general, however, such correlations should be viewed with caution at the present time in the light of these uncertainties, particularly in view of the fact that the validity of these correlations may be dependent on the size of the microstructural unit involved in the fracture process.

7. DYNAMIC INITIATION TOUGHNESS DETERMINATIONS

Since the subject of dynamic toughness in relation to crack arrest forms part of Chapter 6, discussion here will be limited to toughness measurements under dynamic loading conditions relevant, for example, to the assessment of a transient loading situation. The effect on toughness of increasing the speed of load application has been observed to be markedly dependent on the material and the applied strain rate. It is generally anticipated that higher strain rates should lead to a reduction in toughness as a result of the concomitant increase in the resistance of the material to yielding; however, a rise in toughness has been observed in several materials particularly at high loading rates, and it has been suggested that this effect could be due, at least in part, to adiabatic heating in the crack tip region.[52] In the light of such data, it is thus important, if indeed some appreciation of the effect of loading rate is required, to conduct such dynamic tests at a strain rate which is representative of that likely to occur in the structural component of interest.

Clearly, the problems outlined earlier with regard to the detection of first crack extension in static tests are magnified several-fold when the test is conducted at high strain rates. Conventional static test equipment and analysis methods can generally be employed, however, up to loading rates equivalent to a $\dot{K}$ of approximately $3 \times 10^4 \, \text{MNm}^{-3/2}\text{s}^{-1}$ provided that the response times of the test equipment are adequate. Beyond this limit, however, impact machines are generally required which, when suitably modified, enable a load/time record to be established for the test sample. Analysis of such a record is complicated by virtue of the fact that inertial oscillations are set up in the specimen and striker and which are reflected in the load/time trace.[53] The Charpy impact machine has been widely employed as a test facility for conducting dynamic initiation tests although, because of its size, a fatigue precracked Charpy specimen theoretically has a very restricted range of applicability for measuring dynamic plane strain toughness values, K_{Id}: for a dynamic yield strength value of $500\,\mathrm{MNm^{-2}}$, the static specimen size validity requirement on crack length would imply a measuring capacity for the Charpy specimen of up to a K_{Id} value of $22\,\mathrm{MNm^{-3/2}}$. In practice, such specimens are generally employed to measure toughnesses in excess of the strict range of validity in order to obtain a feel for the effect of strain rate on toughness. As a result of the observed effect of test piece size on toughness at static loading rates, the use of larger specimens comparable in terms of thickness and hence constraint to the component of interest is likely to become obligatory and indeed moves in this area are being and have been made (for example, ref. 54). Generally, however, an increase in specimen size leads to a magnification of the inertial effects and the problems of data analysis.

The analysis of brittle failures, whether these occur in the linear elastic regime and thus characterised in terms of K or beyond general yielding in which case J and/or COD analyses are appropriate, can be carried out on the assumption that the maximum recorded load is the relevant measurement point.[55] It is possible, however, that this load value reflects the continued oscillation of the system and thus may not be indicative of the point of fracture, particularly if this maximum occurs in the early part of the test where the specimen is being accelerated.[53] An alternative procedure for the linear elastic range is to instrument the specimen with a conducting paint line immediately beneath the stress concentrator and to employ the time to fracture for the derivation of toughness.[53] The assumptions associated with this method are that the paint line fracture accurately records the point of first crack extension, that crack initiation on the surface and mid-thickness occur simultaneously and that the specimen displacement rate is approximately constant during the test.

In the ductile fracture regime or where slow crack growth is thought to be present, two basic techniques have recently emerged for evaluating dynamic general yielding toughness parameters at high loading rates.[56,57] Both require the testing of several similar samples in order to permit the extrapolation of the resulting resistance curve to zero crack extension. The first method advocates the use of a rigid stop to permit the test to be terminated at a particular deflection.[57] The second uses a modified test piece geometry similar to that detailed in Fig. 7 to enable the specimen to unload automatically at a test piece deflection which depends on the shoulder width, p, to loading span, S, ratio adopted for the particular specimen. By either altering the position of the rigid stop for each sample tested for the former method or by testing a number of similar samples each with a different shoulder width, p, for the latter method, each specimen is thus tested to different bend angles and permits the derivation of a resistance curve. Both methods thus provide off-load δ_i values directly from the measurement of the post-test, off-load angle of deflection and permit the evaluation of J either from a load/time trace by the recommended energy correlation methods (eqn. (3)) or alternatively from the observed off-load

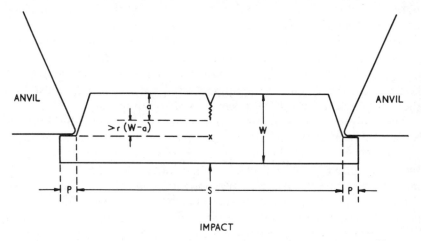

FIG. 7.   Advocated specimen geometry for automatic unloading.[55]

angle of bend by approximate numerical techniques[11] without the need necessarily for measuring load directly during the test. It is emphasised that neither of these techniques has yet been used extensively and both appear to have certain drawbacks. For the former method the basic assumption is that the stop employed to restrict the bending of the test piece is infinitely rigid so that the test piece is subject to an approximately constant deflection rate throughout the test. In the latter method, intense plastic deformation occurs at the shoulders and reduces the deflection rate somewhat from the value provided by the test facility. It is to be borne in mind, in addition, that there is no real theoretical justification for the use of the J integral in unloading situations in general and thus its use in the dynamic regime where oscillatory loading occurs is open to criticism. While this is generally accepted it is hoped, nevertheless, that the J-integral criterion may still be used to characterise approximately the conditions for crack initiation under dynamic conditions to an acceptable accuracy.

While test methods are therefore available for establishing dynamic toughness measurements, which provide some appreciation of the sensitivity of a material to strain rate, the author is not aware of any dynamic analysis that has permitted the application of such high strain rate data to the assessment of defects in structures. Some early instrumented impact data have been considered in the formulation of the K_{Ia} curve concept for the evaluation of flaws under reactor fault conditions, but defect assessment on the basis of the K_{Ia} curve is based on a purely static analysis.[58]

8. TOUGHNESS MEASUREMENTS BEYOND INITIATION

The previous sections have served to highlight the recent developments in techniques for characterising the conditions for first crack extension under essentially static and dynamic loading conditions. Two further areas of basic research, however, are attempting to focus on the possibility of establishing viable measurements of material toughness subsequent to the initiation event: these are namely a study of the conditions under which an originally stable crack growth process becomes unstable and the possibility of characterising the point at which an unstable propagating crack arrests. Discussion on the former aspect is presented in the following paragraphs.

While stable crack extension has been observed under fully plane strain linear elastic conditions,[2] its appearance under such conditions is in fact contrary to energy balance considerations. Such crack growth can be explained, however, either on the basis of the presence of plane stress surface effects or by means of the assumption that the shape of the crack profile alters as the defect extends[59] and recent research has sought to characterise such stable growth in terms of a resistance R-curve approach. While crack stability is maintained the total energy release rate, G, and the total energy absorption rate, R, are equivalent and instability only occurs when G exceeds R beyond the point where

$$\frac{\mathrm{d}G}{\mathrm{d}a} = \frac{\mathrm{d}R}{\mathrm{d}a} \tag{34}$$

The general conclusion is that the precise location of the instability point will be dependent on the precise shape of the resistance R-curve, the flow properties of the material, the geometry under consideration and the type of loading (including system stiffness). Recent research has thus been involved in evaluating the theoretically and experimentally based proposal[60] that the R-curve is independent of test piece geometry. Such work has focused predominantly on the linear–elastic test regime and has resulted in the formulation of a proposal for conducting and analysing R-curve tests.[61] It is necessary in all such tests to have a reliable on-test method of establishing the crack extension occurring as a function of load and compliance techniques are usually employed for this purpose. These techniques for monitoring crack extension provide, at best, estimates of the mean crack growth across the section and the K_R value corresponding to an inferred value of crack growth is then evaluated on the basis of load and the compliance function relevant to the current crack length. Finite element analysis has recently shown, however, that such an averaging procedure in terms of crack length can provide a K_R value in excess of that for the real crack profile which generally tends to exhibit a tunnelling profile at the specimen mid-thickness[59] and it may transpire that when the real profile of a propagating crack is assessed the resistance curve approach may be shown to be a more successful concept than it would currently seem. The present situation with regard to K_R testing is that there are conflicting data both in support of[60] and in conflict with[62] the proposal that the R-curve is geometry invariant. While further work is required to clarify this aspect further, broadly analogous studies are being attempted for more ductile materials using extensions to the toughness concepts described earlier.

It has already been noted in the previous sections that the conditions for crack initiation appear to be well characterised by the COD and J-integral concepts. In situations in which initiation is followed by stable crack extension, as is usually the case in high toughness materials at high temperature, the conditions for initiation can represent considerable underestimates of the conditions for specimen instability.[14−16,33,34] As an example, Table 2 details the results obtained in geometrically similar bend test pieces of 316 stainless steel at 20 °C:[16] it is to be noted that for such

TABLE 2

INITIATION (δ_i) AND MAXIMUM LOAD (δ_m) VALUES OF COD FOR 316 STEEL AT 20 °C[16]
Bend specimen geometry: $B = W = 0.25\,S, \; a = 0.35\,W$

$B\,(mm)$	$\delta_i\;(mm)$ $(\pm 0.01\,mm)$	$\delta_m\;(mm)$ $(\pm 0.02\,mm)$	Δa at maximum load (mm)
10	0.38	0.66	0.64
20	0.38	0.72	0.76
50	0.38	1.10	1.40

geometrically similar specimens the larger the test piece, the larger is the difference between the COD at instability (evaluated at the original crack tip location at maximum load), δ_m, and that at initiation, δ_i. This appears to be a quite general observation for wholly fibrous fractures and has also been noted by Fearnehough et al.[63] and in terms of J measurements by Begley and Landes[14] and Griffis,[15] in so far as the value of δ_m (or J_m) becomes a progressively better but non-conservative estimate of δ_i (or J_i) as the specimen size is reduced towards the less restrictive geometry requirements for plane strain (eqns. (10) and (18)). It is perhaps worthwhile noting in passing therefore that, for deeply notched SEN bend or CT specimens for which the J-integral and equivalent energy concepts become equivalent, K_{IcB} measurements on such specimens are likely to represent upper bound values for K_{Ic} which approach K_{Ic} as the specimen size is *reduced*. This argument of course presupposes that the K_{IcB} and K_{Ic} tests exhibit an identical fracture mechanism and assumes a relationship between K_{Ic} and J_i of the form $K_{Ic}^2 = E'J_i$.

Recent experimental investigations have focused on the possibility of characterising the crack growth resistance of a given material either by using the COD approach or by extending the theoretical range of applicability of the J contour integral. Modifications to the estimation procedures for J (e.g. eqn. (3)) have been proposed as a possible method for charactcrising crack growth but recent R-curve data have not substantiated the previous claim (albeit for the elastic case) that the R-curve may well be a material property;[52] the relationships developed between such pseudo-J measurements and crack extension indicated that the R-curve developed in tension was substantially steeper than that derived in bending.[64] Ductile crack growth, however, appears in some ways to be modelled more effectively by means of crack opening displacement measured at the location of the original crack tip. It has been found that at ductile crack initiation, the first crack advance occurs when a critical strain has been developed at the crack tip, ε_f.[33,42] The further growth of the ductile crack requires an additional tip opening, δ_p, for each microstructural unit of crack advance, s (e.g. the inclusion spacing of the material) sufficient to generate the fracture strain, ε_f, at the new crack tip.[34] This model of fibrous crack growth has been substantiated by Garwood[65] through the use of rubber infiltration moulding of ductile crack tips and has been supported by a linear relationship observed between COD and Δa by several workers.[16,34] Thus:

$$\delta - \delta_i = \Delta a(\delta_p/s) \qquad \delta_i \leq \delta \leq \delta_m \qquad (35)$$

The constancy of the crack tip profile essentially implies that the development of a resistance curve is not related to a requirement for an increased crack tip opening as the crack growth proceeds but rather develops as a result of the loss of constraint brought about by the tunnelling profile adopted by the crack and the requirement to push ahead of the crack the high strain field immediately adjacent to the crack tip and the associated plastic zone. On the basis of this latter requirement, Green and Knott[66] have developed an approximate model of the propagation energy associated with the growth of a ductile crack. Further work has still to substantiate the extent to which the apparently constant profile assumed by a propagating ductile crack is retained in differing specimen geometries. Hopefully, however, an insight into the slow crack growth process and the crack tip conditions which lead to instability may be gained by the consideration of such work terms and elastic energy release.

Despite these ongoing studies, it is still uncertain if and how some allowance for stable tearing could be permitted in structural situations. Application of the R-curve concept would first require a clear demonstration of the uniqueness of such a relationship at least for various loading configurations and some appreciation of the role of stored energy in the instability process. It would appear therefore that as a result of such uncertainties, the toughness parameter to be used for the assessment of structural integrity should currently be strictly limited to one which characterises the initiation event (K_i, δ_i, J_i). While such an approach is consistent with current recommended or proposed practice for COD and J measurements respectively, it does not necessarily comply with the recommended practice for K_{Ic} testing which is such as to permit a crack extension of up to $0.02\,a$,[1,2] where a is the original crack length, at the K_{Ic} measurement point. This essentially means that the K value at the initiation of crack extension, K_i, may in some instances be significantly below the apparently valid K_{Ic} value depending on the slope of the K_R curve. This inconsistency between K_{Ic} practice and current practice for J measurements has been cited as the reason for the fact that J_i has been found in certain cases to be significantly below the value consistent with the valid K_{Ic} data established for the same material on larger specimens.[64] The uncertainties involved in permitting such levels of crack growth essentially imply that the K_{Ic} test specification should be altered to specify the absence of crack extension at the toughness measurement point particularly for specimens which exhibit a rising load trace. Unfortunately, a requirement of this sort would necessitate a multi-specimen test procedure for K_{Ic} similar to that advocated for J. Crack monitoring techniques could prove useful, however, in the linear elastic regime and hence permit the evaluation of K_i from a single specimen test; it is possible, for example, that the alternating current electrical potential method could be particularly useful in this area since the sensitivity is essentially independent of test piece size and for valid tests, the extent of plasticity (and hence the region of increased resistivity) is likely to be less of a problem than in extensively yielded samples.

CONCLUSIONS

The preceding sections have attempted to examine and review recent developments in the methods for assessing material toughness. It would currently appear that while experiments are being conducted in an attempt to identify reproducible and meaningful toughness values beyond that required for the initiation of fracture, these studies have not yet reached a stage at which the conditions for instability are sufficiently well understood. An initiation approach to toughness evaluation and defect assessment procedures would therefore seem to be obligatory for all toughness test methods at the present time as is specified or recommended for the application of the J and COD concepts.

Three static general yielding toughness evaluation procedures have been reviewed and compared and have been shown to be equivalent or related at least for certain test piece geometries and failure conditions. It was noted, however, that a test piece thickness equivalent to that of the structural component of interest was an essential requirement in order to ensure a close approximation of the constraint in service; results have indicated that smaller specimens may exhibit significantly larger toughness values and a different fracture mode when compared to a specimen of the structural thickness particularly in the temperature transition range.

The possibility of being able to predict static toughness values from Charpy and uniaxial tensile tests has been discussed and while such approaches may well prove to be widely applicable, subject to further confirmation, it would appear that the correlations derived may well be additionally dependent on the size of microstructural unit relevant to the failure mechanism (e.g. grain size or inclusion spacing).

REFERENCES

1. BRITISH STANDARDS INSTITUTION. 'Methods of test for plane strain fracture toughness (K_{Ic}) of metallic materials'. BS 5447: 1977.

2. AMERICAN SOCIETY FOR TESTING AND MATERIALS. 'Standard method of test for plane-strain fracture toughness of metallic materials'. E399-72, 1972.
3. LARSSON, S. G. and CARLSSON, A. J. *J. Mech. Phys. Solids*, 1973, **21**, 263.
4. RICE, J. R. *J. Appl. Mech.*, *Trans. ASME*, 1968, **35**, 379.
5. RICE, J. R. In *Fracture—An Advanced Treatise*, Ed. H. Liebowitz, 1968, Academic Press, New York.
6. BEGLEY, J. A. and LANDES, J. D. 'Fracture toughness', 1972, *ASTM STP* 514, 1.
7. LANDES, J. D. and BEGLEY, J. A. 'Fracture toughness', 1972, *ASTM STP* 514, 24.
8. BOYLE, E. F. 'Calculation of elastic and plastic crack extension forces', PhD Thesis, Queen's University, Belfast, 1972.
9. RICE, J. R., PARIS, P. C. and MERKLE, J. G. 'Progress in flaw growth and fracture toughness testing', *ASTM STP* 536, 231.
10. SUMPTER, J. D. G. and TURNER, C. E. 'Cracks and fracture', 1976, *ASTM STP* 601, 3.
11. BUCCI, R. J., PARIS, P. C., LANDES, J. D. and RICE, J. R. 'Fracture toughness', 1971, *ASTM STP* 514, 40.
12. MERKLE, J. G. and CORTEN, H. T. *J. Press. Vessel Techn.*, *Trans. ASME*, 1974, **12**, 286.
13. CHIPPERFIELD, C. G. A summary and comparison of *J* estimation procedures. *J Test. and Evaluation*, 1978, **6**, 253.
14. LANDES, J. D. and BEGLEY, J. A. Recent developments in J_{Ic} testing. Westinghouse Research Laboratories, Pittsburgh, Pennsylvania Scientific Paper 76-1E7-JINTF-P3, 1976.
15. GRIFFIS, C. A. *J. Press. Vessel Techn.*, *Trans. ASME*, 1975, **13**, 278.
16. CHIPPERFIELD, C. G. *Int. J. Fracture*, 1976, **12**, 873.
17. BURDEKIN, F. M. and STONE, D. E. W. *J. Strain Anal.*, 1966, **12**, 145.
18. LOGSDON, W. A. 'Mechanics of crack growth', 1975, *ASTM STP* 590, 43.
19. RICE, J. R. *J. Mech. Phys. Solids*, 1974, **22**, 17.
20. HOPKINS, P. and JOLLEY, G. In *Fracture 1977*, vol. 3, Ed. D. M. R. Taplin, 1977, 329.
21. SUMPTER, J. D. G. and TURNER, C. E. *Int. J. Fracture*, 1976, **12**, 887.
22. BEGLEY, J. A., LANDES, J. D. and WILSON, W. K. 'Fracture analysis', 1974, *ASTM STP* 560, 155.
23. BURDEKIN, F. M. The British Standard Committee WEE/37 Draft and IIW Approaches. This volume, Chapter 3.
24. WITT, F. J. Some observations on size effects in impact energy and their implications to fracture behaviour. Fourth Annual Information Meeting for the HSST Programme, 1970.
25. WITT, F. J. Equivalent energy procedures for predicting gross plastic fracture. Fourth National Symposium on Fracture Mechanics, Carnegie-Mellon University, 1970.
26. WITT, F. J. and MAGER, T. R. A procedure for determining bounding values on fracture toughness, K_{Ic}, at any temperature. Fifth National Symposium on Fracture Mechanics, University of Illinois, 1971.
27. BUCHALET, C. and MAGER, T. R. 'Fracture toughness testing', 1975, *ASTM STP* 536, 281.
28. WITT, F. J. and MAGER, T. R. *Nucl. Eng. and Design*, 1971, **17**, 91.

29. CHELL, G. G. An assessment of some methods for obtaining fracture toughness values from invalid test data. CEGB Report RD/L/N 18/76, 1976, Central Electricity Research Laboratories, UK.
30. WELLS, A. A. Symposium on Crack Propagation. College of Aeronautics, Cranfield, 1961, Paper B4.
31. COTTRELL, A. H. ISI Spec. Report No. 69, 1961, 281.
32. BRITISH STANDARDS INSTITUTION. 'Methods of test for Crack Opening Displacement testing'. Draft for Development, DD19, 1972.
33. KNOTT, J. F. *'Fundamentals of Fracture Mechanics'*, 1973, Butterworth, London.
34. GREEN, G. G. and KNOTT, J. F. *J. Mech. Phys. Solids.*, 1975, **23**, 167.
35. ROBINSON, J. N. and TETELMAN, A. S. *Int. J. Fracture*, 1975, **11**, 453.
36. HANCOCK, J. W. and MACKENZIE, A. C. *J. Mech. Phys. Solids*, 1976, **24**, 147.
37. HAYES, D. J. and TURNER, C. E. *Int. J. Fracture*, 1974, **10**, 17.
38. GREEN, A. P. and HUNDY, B. B. *J. Mech. Phys. Solids*, 1956, **4**, 128.
39. SUMPTER, J. D. G. 'Elastic–plastic fracture analysis and design using the finite element method'; PhD Thesis, London University, 1974.
40. ROBINSON, J. N. *Int. J. Fracture*, 1976, **12**, 723.
41. CLARKE, G. A., ANDREWS, W. R., PARIS, P. C. and SCHMIDT, D. W. 'Mechanics of crack growth', 1975, *ASTM STP 590*, 27.
42. CHIPPERFIELD, C. G. and KNOTT, J. F. *Metal Techn.*, 1975, **2**, 45.
43. BARSOM, J. M. and ROLFE, S. T. 'Impact testing of metals', 1970, *ASTM STP 466*, 27.
44. SAILORS, R. H. and CORTEN, H. T. 'Fracture toughness', 1971, *ASTM STP 514*, 164.
45. MARANDET, B. and SANZ, G. Evaluation of the toughness of thick medium strength steels by using linear elastic fracture mechanics and correlations between K_{Ic} and C_v. Irsid Institut de Recherches de la Siderurgie Francaise, 1976, P254.
46. WILSHAW, T. R., RAU, C. A. and TETELMAN, A. S. *Eng. Fract. Mech.*, 1968, **1**, 191.
47. RITCHIE, R. Q., FRANCIS, B. and SERVER, W. L. *Met. Trans.*, 1976, **7A**, 831.
48. ROLFE, S. T. and NOVAK, S. R. 'Review of developments in plane strain fracture toughness testing', Ed. W. F. Brown, jnr., 1970, *ASTM STP 463*, 25.
49. COPELAND, J. F., YUEN, J. L. and OFFER, H. P. British Nuclear Energy Society Conference on Ferritic Steels for Fast Reactor Steam Generators, London, 1977, Paper 23.
50. CHIPPERFIELD, C. G. *Preprints of the Institution of Mechanical Engineers Conference on Tolerance of Flaws in Pressurised Components.* Conference to be held in London, May 1978.
51. KNOTT, J. F. Institute of Physics and Metals Society Conference on the Mechanics and Physics of Fracture. Cambridge University, 1975, Paper 9.
52. KRAFFT, J. M. and IRWIN, G. R. 'Fracture toughness testing and its applications', 1968, *ASTM STP 381*, 110.
53. TURNER, C. E., CULVER, L. E., RADON, J. C. and KENNISH, P. Institution of Mechanical Engineers Conference on Practical Application of Fracture Mechanics to Pressure Vessel Technology. London, 1971, Paper C6/71.
54. IIW Commission X, UK Briefing Group on Dynamic Testing. The Welding

Institute and American Society for Metals Conference on Dynamic Fracture Toughness. London, 1976, Paper 11.

55. IRELAND, D. R., SERVER, W. L. and WULLAERT, R. A. 'Testing and data analysis procedures', Effects Technology Incorporated Technical Report 74-32, 1975 (revised).

56. CHIPPERFIELD, C. G. The Welding Institute and American Society for Metals Conference on Dynamic Fracture Toughness. London, 1976, Paper 16.

57. WULLAERT, R. A. and SERVER, W. L. Fracture Control Corporation, Goleta, California. Private communication, 1976.

58. COWAN, A. The Approach to Analysis of Significance of Flaws in ASME Section III and Section XI. This volume, Chapter 2.

59. NEALE, B. K. *Int. J. Fracture*, 1976, **12**, 499.

60. KRAFFT, J. M., SULLIVAN, A. M. and BOYLE, R. W. *Proc. of the Symposium on Crack Propagation*, 1961, **1**, 8. College of Aeronautics, Cranfield.

61. AMERICAN SOCIETY FOR TESTING AND MATERIALS, *Annual Handbook of Standards*. 1974 (part 10).

62. BRADSHAW, F. J. and WHEELER, C. The crack resistance of some aluminium alloys and the prediction of thin section failure. Royal Aircraft Establishment Technical Report 73191, 1974.

63. FEARNEHOUGH, G. D., LEES, G. M., LOWES, J. M. and WEINER, R. T. Institution of Mechanical Engineers Conference on Practical Application of Fracture Mechanics to Pressure Vessel Technology. London, 1971, Paper 33/71.

64. BEGLEY, J. A. and LANDES, J. D. *Int. J Fracture*, 1976, **12**, 764.

65. GARWOOD, S. J. The measurement of crack growth resistance using yielding fracture mechanics. PhD Thesis, Imperial College of Science and Technology, London, 1976.

66. GREEN, G. and KNOTT, J. F. American Society of Mechanical Engineers Conference on Micromechanical Modelling of Flow and Fracture. Troy, New York, 1975, Paper 75-Mat-10.

67. MOHAMED, T. and TETELMAN, A. S. *Eng. Fract. Mech.*, 1975, **7**, 107.

68. INGHAM, T. Risley Nuclear Power Development Laboratories, UKAEA, Risley, Warrington, Cheshire. Private communication.

69. SMITH, R. F. and KNOTT, J. F. Institution of Mechanical Engineers Conference on Practical Application of Fracture Mechanics to Pressure Vessel Technology. London, 1971, Paper C9/71.

Chapter 6

DYNAMIC CRACK PROPAGATION AND ARREST

D. Francois

Université de Technologie de Compiègne, France

SUMMARY

This chapter, which deals with the behaviour of dynamic crack propagation and arrest, begins with a description of the stress field singularity at a running crack tip. It depends on a dynamic stress intensity factor, which is a function of the wave velocities. Various experimental verifications are mentioned. Conditions for crack arrest are then stated and the experiments for the study of this phenomenon are described. Finally, some experimental results on non-metallic as well as on metallic materials are given.

1. INTRODUCTION

The evaluation of the fracture resistance of welded structures has for many years been considered through the use of Charpy tests, which provide a qualitative comparison of metals by the indication of the safety margin given by the difference between the transition temperature and the lowest service temperature. More recently, fracture mechanics has brought a quantitative relation between the stresses in the weldment, the size of the defects which might occur there and the fracture toughness. Hopefully it becomes then possible to achieve a safe analysis of welded structures. Some difficulties are encountered concerning the residual stresses which must be

151

added to the applied stresses in the calculations, uncertainties about the true defect sizes and the measurement of the fracture toughness in a material where the welding operation has produced large gradients of the mechanical properties. It is thus difficult to guarantee, in some cases where safety is paramount, that fracture will not occur at all. Furthermore it is most probable that the zones where there are defects are also those where structural anomalies are likely to be met for which no value of the fracture toughness would be available. This is the reason for the consideration of an alternative safety approach based on ascertaining that any such crack would arrest in the base metal. The risk of initiation in the weldment would not be eliminated, but the conditions would be so defined that a crack, even though it would have acquired a large velocity, would arrest in the base metal. For this reason several laboratories are studying the conditions which will cause the arrest of fast running cracks. However, those experiments meet with some difficulties.

The first one is to know the right mechanical parameter to use. As seen in the earlier chapters, under certain conditions in the case of statically loaded cracks, the crack extension force, G, related to the stress intensity factor, K, is the right parameter (if the plastic deformation is limited small-scale yielding approximation). In a dynamic case inertia effects perturb the stress field near the tip of the running crack, the deformations being propagated from one point to the next as waves. It is relatively easy to calculate the behaviour in a large structure, but reflections of the waves on the side surfaces lead to a great complication in a piece of finite dimensions such as a test specimen. On the other hand, part of the energy of the system is effective by kinetic energy which raises a problem in relation to when the crack comes to a stop as to what happens to this energy. Is it entirely available for the propagation or is it dissipated through other non-reversible mechanisms? To those theoretical difficulties must be added the ones which are connected with tests of very short duration and which require a complex instrumentation.

However, some clear lines are emerging. After Erdogan's review[1] and the 1972 meeting at Lehigh University,[2] more recent conferences[3,4] have given an overall view on the subject. This chapter is an attempt to gather the main findings. It first deals with the changes in the stress field at the crack tip and thus in K which are introduced in dynamic cases. It describes the main experimental techniques used to study crack arrest and gives some results. Thus although it cannot provide the firm basis for the introduction of crack arrest concepts or dynamic loading effects into flaw tolerance analyses, it should indicate the difficulties and future possibilities.

2. THE STRESS FIELD AT A RUNNING CRACK TIP, K_I^{dyn}

2.1. Theoretical Calculations

A crack whose tip is running at velocity V is considered. The theory of elasticity[5−7] shows that the stress field near the tip is given by:

$$\sigma_{ij} = K_1(2\pi r)^{-1/2} F_1(S_1, S_2) f_{ij}(\theta, S_1, S_2) \tag{1}$$

where S_1 and S_2 are functions of the speeds of the dilatation waves C_1 and the shear waves C_2, and of the crack velocity.

$$S_1^2 = I - (V/C_1)^2$$
$$S_2^2 = I - (V/C_2)^2 \tag{2}$$

with

$$C_1 = (\lambda + 2\mu)^{1/2}/\rho^{1/2} \quad \text{and} \quad C_2 = (\mu/\rho)^{1/2} \tag{3}$$

μ and λ being the Lamé constants and ρ the density.

Thus the stress field possesses as in the static case a $r^{-1/2}$ singularity and it can be shown that the dynamic stress intensity factor K_I^{dyn} is related to the static K_1 by the equation:

$$K_I^{dyn} = K_1 F(S_1, S_2, \text{geometry}) \tag{4}$$

However, some authors[26] think that this relation should also incorporate the crack acceleration and that there might be two branches in the $K_I^{dyn}(V)$ relationship, one for accelerating and one for decelerating cracks as shown by some results of Kobayashi and Dally.[27]

The function F depends upon the geometry of the particular system which is considered and unhappily is not yet known in all cases, further numerical analyses being required. It has, however, been determined in a few simple cases, of which the following are examples.

Yoffe's Solution[8]
Crack of constant length $2a$ propagating at constant speed in an infinite plate with a homogeneous applied tensile stress. In this unrealistic case where one tip closes as the other opens, $K_I^{dyn} = K_I$.

Broberg's Solution[9,10]
Crack of length $2a$ which opens at both ends with a constant speed in an infinite plate with a homogeneous applied tensile stress. K_I^{dyn} decreases almost as $V^{-1/2}$ and becomes zero when the crack velocity reaches the Rayleigh wave speed (Fig. 1).

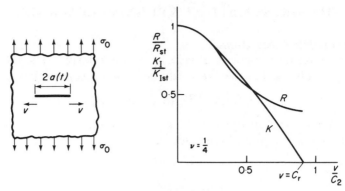

FIG. 1.   Broberg's solution for a running crack of length $2a$ which opens on both sides at a constant velocity v in an infinite plate.

Craggs' Solution[11]

Semi-infinite crack in an infinite plate. The solution is analogous to the preceding one. This problem was, in particular, worked out by Freund.[12,13]

Nilsson's Solution[14,15]

Semi-infinite crack propagating at constant velocity in a narrow plate of height h with a constant applied displacement.

As in the preceding cases K_I^{dyn} decreases and reaches zero when the crack velocity approaches the Rayleigh wave velocity (Fig. 2) the difference with Broberg's and Craggs' solutions shows the influence of the sides where the waves are reflected: the drop in K_I^{dyn} is then much slower than in an infinite plate.

DCB Specimens

These are often used in crack arrest studies as the design provides a K factor which decreases as the length of the crack increases. Calculations were performed at the Battelle Memorial Institute by Hahn et al.[16] They use for this purpose the dynamic behaviour of a beam on an elastic foundation. Comparison of the obtained results with those of Freund[12] clearly shows the influence of the reflections of the waves on the sides of the specimen.

The spatial distribution of the stresses at the crack tip are given by the function $f_{ij}(\theta, S_1, S_2)$ in formula (1). It is independent of geometry but it varies with the crack velocity as shown on Fig. 3. After a velocity of about $0.7C_2$ is reached the $\sigma_{\theta\theta}$ stress does not reach its maximum value anymore

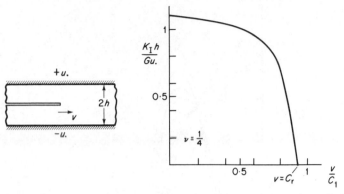

FIG. 2. Nilsson's solution for a semi infinite running crack in a plate of width $2h$ with a displacement u on the sides.

Plane stress:

$$K_I^{dyn} = K_I[4S_1S_2 - (1 + S_2^2)^2]^{1/2}/[S_1(1 - S_2^2)]^{1/2}(1 + v)^{1/2}$$

Plane strain:

$$K_I^{dyn} = K_I(1 - v)^{1/2}[4S_1S_2 - (1 + S_2^2)^2]^{1/2}/[S_1(1 - S_2^2)]^{1/2}$$

with

$$K_I = uE/[h(1 - v^2)]^{1/2} \quad \text{Plane stress}$$
$$K_I = uE/[h(1 + v)^2(1 - 2v)]^{1/2} \quad \text{Plane strain}$$

for $\theta = 0$, i.e. in the crack propagation plane, but for $\theta \simeq 60°$. This could explain why above a critical velocity the crack shows a tendency to bifurcate and to branch.

Another important remark is concerned with the evolution of the stress triaxiality. In the propagation plane for instance ($\theta = 0$)

$$\sigma_{22}/\sigma_{11} = [4S_1S_2 - (1 + S_2^2)^2]/[(1 + 2S_1^2 - S_2^2)(1 + S_2^2) - 4S_1S_2] \quad (7)$$

This stress ratio decreases as the crack velocity increases, from unity for $v = 0$ to zero when v reaches the Rayleigh wave velocity. Thus the shear stress in this plane increases which makes the plastic deformation easier. The evolution of the density of elastic energy shows the same effect. So an increase of the fracture energy would be expected when the crack velocity increases.

Few studies take into account the plastic deformation. A paper by Embley and Sih[17] using the Dugdale approach and one by Broberg[18] using a plastic sheet approach are to be mentioned.

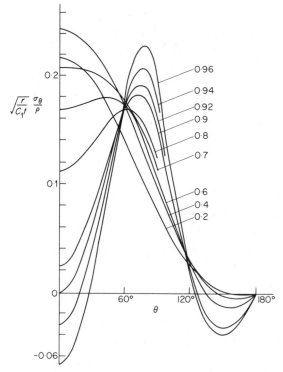

FIG. 3. Variation of the stress σ_θ as a function of the polar angle at the crack tip for various values of the crack velocity (with respect to the shear wave velocity).[1]

It is remarkable that the preceding results show a striking analogy with those obtained on dislocations (see for instance *Theory of dislocations*, J. P. Hirth and J. Lothe, McGraw-Hill, New York (1968)).

A good understanding of the dynamic crack propagation and arrest is provided by an energy balance analysis. The first thermodynamics principle, under adiabatic conditions, is written:

$$\frac{\partial}{\partial a}(W - U - T) = \frac{\partial D}{\partial a} \tag{8}$$

where W is the energy provided to the system, U the deformation energy, T the kinetic energy and D the energy which is absorbed irreversibly as the crack length a is being increased. D can be looked upon as the sum of the fracture energy, γ, and of the energy spent in various dissipative processes, D_p.

The notion of the crack extension force can be generalised in the dynamic case as

$$G^{dyn} = \frac{\partial}{\partial a}(W - U - T)$$

It can be shown that G^{dyn} is related to G by the relation

$$G^{dyn}/G = [1/(1 - v)]S_1(1 - S_2^2)[4S_1S_2 - (1 + S_2^2)^2]F_1^2(S_1S_2) \qquad (9)$$

which is to be compared with relation (4). This ratio thus varies in the same fashion as K_I^{dyn}/K_I.

2.2. Experimental Verifications

In order to verify these predictions, techniques must be used which allow the deformation field at the running crack tip and its evolution to be analysed as the crack propagates. Essentially three techniques have been used: strain gauge extensometry, photoelasticity and caustic method.

Strain Gauge Extensometry
Klemm and Shönert used very small rosettes which were fabricated by evaporation near the crack tip in a glass plate.[19] The results were in good agreement with Broberg's.

Dynamic Photoelasticity
This method has been mainly used by Kobayashi.[20] The principle is the same as in static photoelasticity. A photoelastic material must be used and the isochromatics obtained with a polarised monochromatic light source are filmed with a high-speed camera. Results were obtained in specimens with an edge notch (SEN), on bend specimens (DTT) and on double cantilever beam specimens (DCB). Comparisons were also made with finite element analysis. As a whole, K_I^{dyn} displayed fluctuations to be related with the reflections of the waves on the sides.

The Caustic Method
This method was introduced by Manogg[21] and was developed by Theocaris.[22] It was used to follow the fast propagation of cracks by Kalthoff, Winkler and Beinert.[23] It consists in the observation of the shadow pattern which is produced by the crossing by a parallel light beam of deformed transparent material. As the deformation changes the thickness of the plate as well as its refraction coefficient, it deviates the path of the light beam. At the tip of a crack the deformation produces a circular

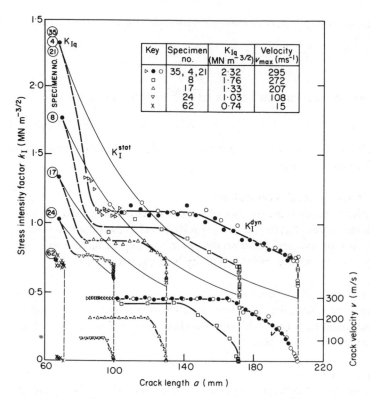

FIG. 4. Results of K_I^{dyn} obtained with the caustic method[23] the initial stress intensity factor, K_{Iq}, is modified by changing the radius of curvature of the notch. The crack velocity is also measured.

caustic whose diameter is proportional to $K^{2/5}$. By filming this caustic with a high-speed camera it is possible to follow the evolution of the stress intensity factor, K, with the crack velocity. Figures 4 and 5 show the results which were obtained on a DCB specimen. This last figure displays quite clearly the influence of the reflections of the waves on the sides of the specimen as they are not in agreement with the calculations on infinite plates: they would provide a unique value of K_I^{dyn}/K_I for each crack velocity and it would keep decreasing. Those results are in good agreement with the calculations of Hahn *et al.*[16] Figure 6 also shows the effect of the reflections. It represents the oscillations of the stress intensity factor around its static value after the crack arrest. In an infinite plate those oscillations would not exist.[12]

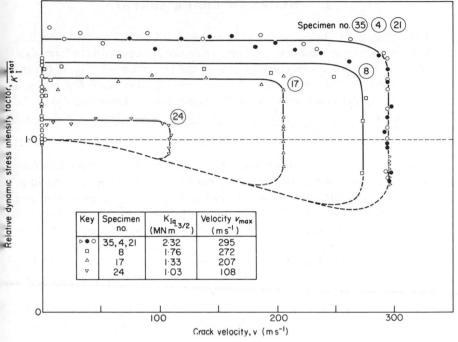

FIG. 5.   The results plotted on the preceding figure are given as a function of the crack velocity. The dashed lines give the assumed evolution at the beginning of the crack velocity where the measurements were not performed.

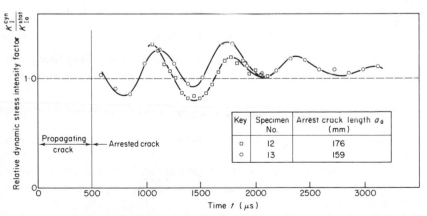

FIG. 6.   Oscillations of the stress intensity factor around the static value after crack arrest. They are due to the reflections of the waves on the sides of the specimen.[23]

3.   THEORY OF CRACK ARREST

The relation (8) helps to build the theory of crack arrest. The first hypothesis is to assume that for a given structure it is possible to formulate G^{dyn} as a function of the crack length, a, and the crack velocity, v. On the other hand, the fracture energy, D, in relation (8), also called R_{ID}, is also a function of the crack velocity, v. It should take a value satisfying the equilibrium

$$G_I^{dyn} = R_{ID} \tag{10}$$

As in the static case, those relations, based on a linear elastic analysis, need the plastic zone to remain small (small-scale yielding approximation). Dahlberg and Nilsson[24] discussed this hypothesis. They showed that if it was verified, the part D_p of the fracture energy due to dissipative processes was a function of v and of K_I^{dyn}. Relation (8) then gives a unique solution for K_{ID}, and thus for R_{ID}, for each velocity v. These authors insist, nevertheless, on the fact that the conditions needed for the small-scale yielding approximation to hold were not easy to put down and that this point required some deeper investigations.

It is the usual hypothesis (which in several cases has been verified by experiments) that K_{ID} goes through a minimum K_{Im} for a particular velocity v.

Crosley and Ripling[25] then assume that a crack stops in an infinite plate when K_I takes a critical value for arrest K_{Ia}, where K_I is the stress intensity factor under static conditions. It all happens as it does at crack initiation as if time was reversed. These authors write eqn. (4) as

$$K_I^{dyn}(a, v) = k(v)K_I(a) \tag{11}$$

It is then clear that when $v = 0$ eqn. (8) leads to $K_I = K_{Ia}$. They then go on stating that if the crack velocity is stabilised prior to crack arrest at a value v_m, just before the crack stops

$$k(v_m)K_I(a) = K_{ID} \tag{12}$$

and thus $K_{Ia} = K_{ID}/k(v_m)$ is a reproducible value which is characteristic of a given material.

Crosley and Ripling add that this hypothesis still holds in a structure of finite dimensions since the damping of the waves should greatly weaken the reflections on the sides.

Hahn et al.,[16] on the contrary, believe that in such a structure at first the crack accelerates and that kinetic energy is stored in the system. It is later

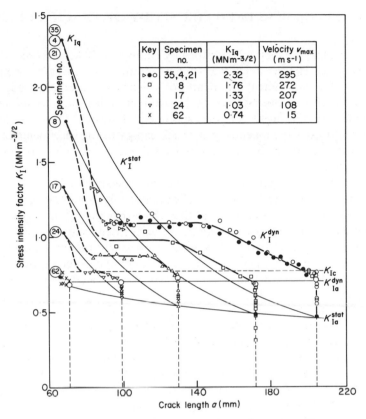

FIG. 7. Measurements of K_{Ia}^{dyn} in Araldite-B.[23]

available to increase the crack extension force at the time of arrest. They also consider that damping does not play a large role. Figure 7, which is due to Kalthoff, Beinert and Winkler,[23] is very demonstrative. It shows that the arrest happens for a value of K_I^{dyn} which seems to be characteristic of the material, whereas the corresponding static K_{Ia} decreases when the initial K_I (and thus the velocity) is increased. However, since it can be argued that the results shown in Fig. 7 may be specific features of particular materials or of the DCB specimen geometry used in those tests, it is desirable that similar experiments be done on other materials and other geometries. Hahn *et al.* consider that the critical value K_{Ia}^{dyn} should be practically equal to K_{Im}, the minimum in the dynamic toughness. This should make the determination of K_{Ia}^{dyn} easier.

4. SPECIMENS USED FOR STUDY OF CRACK ARREST

In recent years the method described by Hahn *et al.*[16] has become the one mostly used. It uses a DCB specimen (or in some cases the compact tension—CT—specimen) the opening of which is produced by a wedge. It is not impossible to open the crack by a direct action of a tensile testing machine on the arms of the specimen. However, this introduces complications as the machine gives back energy to the specimen at arrest. With a wedge, on the other hand, this phenomenon is, in practical terms, non-existent.

In order to propagate the crack at various speeds, notches with various radii of curvature are being used. The greater this radius the larger is the available energy at initiation, and thus the larger the crack velocity. The measurement of the crack velocity can be performed using a resistance grid on the crack path. The arms of the grid are broken as the crack propagates. Experimental measurements have demonstrated that in such specimens the crack accelerates such that it very quickly reaches a stable velocity. The deceleration or arrest is also very fast.

In the dynamic crack propagation experiments the crack has a strong tendency to deviate or to branch. A good way to get rid of this is to provide side grooves on the specimen. As for static tests the question of the size of the specimen to meet the requirements of the small-scale yielding approximation must be considered. A good way to reduce the dimensions is to use a duplex specimen. The front part of such a specimen will include the notch, and be made of a brittle material such that the plastic zone is small. The second or back part, welded to the front one, is made of the tested material. It is in this second part that the crack stops and the plastic zone size is reduced because the elastic limit is raised at high deformation rates.

A variation on this technique described by Nilsson[14] is to use narrow band specimens with a dissymmetric loading.

5. EXPERIMENTAL RESULTS FOR DIFFERENT MATERIALS

5.1. Non-metallic Materials

There are several results for transparent materials since measurements can be made more easily on these. Figure 7 is a good example and shows the behaviour of Araldite-B. K_{Ia}^{dyn} seems to be a little smaller than K_{Ic}, but

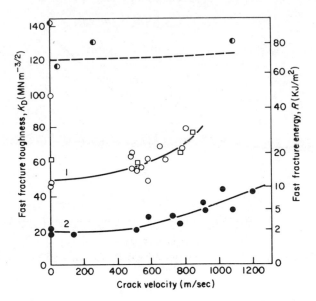

FIG. 8.   Variation of K_D with the crack velocity (1) in type 4340 steel (2) in a 0·1 % C tool steel and (3) (dashed line) in a 9 % Ni steel at $-196\,°C$ (ductile fracture).[16]

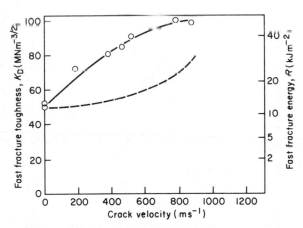

FIG. 9.   Comparison between the variation of K_D with the crack velocity for a fracture with shear lip formation (full line—type 4340 steel 205 °C tempered) and for a ductile flat fracture (dashed line—type 4340 steel 370 °C tempered).[16]

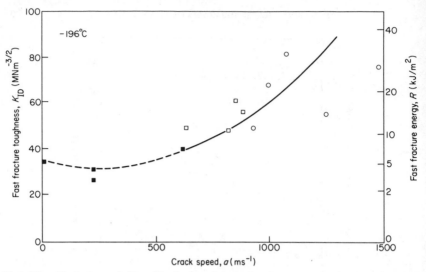

FIG. 10. Variation of K_D with the crack velocity for an ASTM-A517F steel at $-196°C$.[16]

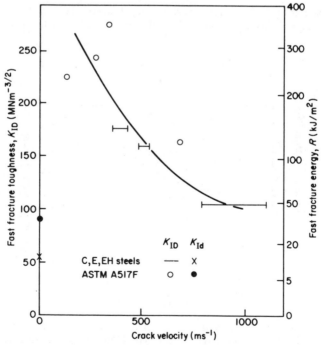

FIG. 11. Variation of K_D with the crack velocity for steels tested near the NDT.[16]

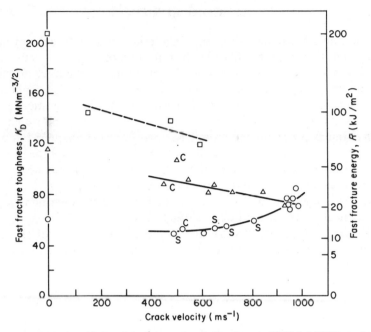

FIG. 12. Variation of K_D with the crack velocity for an ASTM-A533B steel tested at various temperatures: □ +24 °C; △ −12 °C; ○ −78 °C (NDT = −12 °C).[16]

within the experimental scatter. Results by Kobayashi and Mall[20] for Homalite-100, give $G_{Ia}^{dyn}/G_c \simeq 0.5$.

Crosley and Ripling[25] give values of G_{Ia} which range between 35 and 58 J/m^2 for a joint between two aluminium alloy 2024 plates made with epoxide Dow 332.

5.2. Metallic Materials

The most complete results are those of Hahn et al.;[16] they are reproduced in Figs. 8 to 12. They state that the microscopic fracture processes play a dominant role in the dynamic propagation of cracks and thus for arrest as well. The ductile fractures are connected with a dynamic fracture toughness, K_D, which increases weakly with the crack velocity (Fig. 8). If the fracture induces shear lips, their width increases with the crack velocity thus raising K_D (Fig. 9). In the brittle ductile transition, on the other hand, K_D has a low value and it decreases when the crack velocity increases. In this case K_{Ia} should be quite low.

6. CONCLUSION

The measurements to date show that in a finite size structure the dynamic stress intensity factor is modified by the reflection of the waves on the sides. Kinetic energy is stored and it brings its contribution to the crack extension force at arrest. A critical value of the fracture toughness K_{Ia}^{dyn} could exist, this being representative of the material in a particular condition.

An assessment of flaw tolerance under crack arrest condition would then involve similar procedures to those described in earlier chapters but using the appropriate value of K_{Ia}^{dyn} for the bulk material instead of the local K_{Ic} value. In many cases it may be that K_{Ia}^{dyn} may be approximated by the minimum value of the dynamic fracture toughness K_{ID}.

Experimental methods are being studied to ease the determination of these critical values of the fracture toughness.

The first results on steels show that K_{ID} increases weakly with the crack velocity in the ductile region and, on the contrary, decreases strongly near the NDT.

REFERENCES

1. ERDOGAN, F. In *Fracture*, **2** (1968), Academic Press, 498.
2. Dynamic Crack propagation, G. C. Sih (Ed.), Noordhoff (1973).
3. Symposium: 'Fast fracture and crack arrest'. ASTM, Chicago (June 1976), *ASTM STP* 627.
4. Conference: 'Dynamic fracture toughness'. Welding Inst./Am. Soc. Met., London (July 1976). Proceedings published by Welding Inst., London.
5. SIH, G. C. *Int. J. Fract. Mech.*, **4** (1968), 51.
6. RICE, J. R. In *Fracture* **2** (1968), Academic Press, 235.
7. GROSS, D. Advanced Seminar on Fracture Mechanics. Ispra (1975). Proceedings published by Euratom.
8. YOFFE, E. H. *Phil. Mag.*, **42** (1951), 739.
9. BROBERG, K. B. *Arkiv. för Fysik.*, **18** (1960), 159.
10. MANSINHA, L. *J. Mech. Phys. Solids*, **12** (1964), 353.
11. CRAGGS, J. W. *J. Mech. Phys. Solids*, **8** (1960), 66.
12. FREUND, L. B. *J. Mech. Phys. Solids*, **20** (1972), 129; **20** (1972), 141; and **21** (1973), 47.
13. FREUND, L. B. and RICE, J. R. *Int. J. Solids Structures*, **10** (1974), 411.
14. NILSSON, F. *Int. J. Fract. Mech.*, **8** (1972), 403.
15. KUHN, G. and MATCZYNSKI, M. *Eng. Trans. Polska Akad. Nauk.*, **22** (1974), 469.
16. HAHN, G. T. HOAGLAND, R. G., and ROSENFIELD, A. R. *ASTM STP* 627 (1976), and HAHN, G. T., GEHLEN, P. C., HOAGLAND, R. G., MARSCHALL, C. W., KANNINEN, M. F., POPELARD, C. and ROSENFIELD, A. R. BMI. NOREG 1959, Battelle Columbus (1976).

17. EMBLEY, G. T. and SIH, G. C. *Eng. Fracture Mech.*, **4** (1972), 431.
18. BROMBERG, K. B. In *Proc. Int. Conf. on Dynamic Crack Propagation*, Noordhoff (1973).
19. KERKHOF, F. Advanced Seminar on Fracture Mechanics. Euratom, Ispra (1975).
20. BRADLEY, W. R. and KOBAYASHI, A. S., *Eng. Fract. Mech.*, **3** (1971), 317; and KOBAYASHI, A. S. and MALL, S. In *ASTM STP* 627 (1976).
21. MANOGG, P. *Proc. Int. Conference on The Physics of Non-Crystalline Solids*, Delft (1964) 481.
22. THEOCARIS, P. S. *J. Mech. Phys. Solids*, **20** (1972), 265.
23. KALTHOFF, J. F., WINKLER, S. and BEINERT, J. *Int. J. Fract.*, **12** (1976), 317, and KALTHOFF, J. F., BEINERT, J. and WINKLER, S. In *ASTM STP* 627 (1976).
24. DAHLBERG, L. and NILSSON, F. In *ASTM STP* 627 (1976).
25. CROSLEY, P. B. and RIPLING, E. J. In *ASTM STP* 627 (1976).
26. KALTHOFF, J. F. Private communication.
27. KOBAYASHI, T. and DALLY, J. W. The relation between crack velocity and the stress intensity factor in birefringent polymers, Report, Mechanical Engineering Department, University of Maryland (May 1976).

Chapter 7

RECENT DEVELOPMENTS IN FATIGUE CRACK GROWTH ASSESSMENT

I. L. MOGFORD

Central Electricity Generating Board, Leatherhead, UK.

SUMMARY

Fatigue crack growth analyses are equally applicable to assessment of defects actually detected in plant and to design studies of the ability of a proposed structure to tolerate a defect throughout its lifetime of cyclic duty. A brief description of the mechanisms responsible for crack extension under reversed plasticity leads to quantitative crack growth 'laws' in terms of the stress intensity factor amplitude ΔK.

These fatigue crack growth laws are used to assess the cyclic design life of pressure vessels. Crack growth can be calculated using iterative computation or by integration under particular conditions. The uncertainties inherent in complex life analyses are reviewed. For example, fatigue crack growth can be non-cumulative and retardation of cracking can result from non-uniform stress cycling. The threshold stress intensity above which crack growth occurs is particularly sensitive to the mean stress in the cycle. Behaviour at short crack lengths is important since the threshold may be less than that measured experimentally at larger crack lengths; further, at regions of stress concentration a small crack will grow in the enhanced stress field of the concentration.

Fatigue crack growth can be significantly accelerated by common environments like water. The most important parameters influencing the acceleration are mean stress and strain rate during the loading part of the cycle (rather than the cycle frequency).

1. THE NEED FOR FATIGUE ASSESSMENTS

The principle is simple. Any flaw in a pressurised component must not be permitted to grow by a fatigue mechanism to a size at which the component will fail before the end of its design life. This applies to flaws which are detected at any stage in the life of a component and to a postulated flaw being considered during component design analysis.

The normal duty expected of a pressurised component like a gas container, a nuclear pressure vessel, cryogenic tanks, submarines, aircraft cabins or whatever, involves repeated emptying and filling or variation in ambient pressure which impose a cyclic variation in pressure stress. Frequently the fluid is not at room temperature so that thermal gradients can arise in the structure with associated cyclic thermal stresses which are superimposed on cyclic pressure stresses. These 'duty' cyclic stresses can cause stable fatigue extension of pre-existing flaws in the structure and the amount of such extension can be calculated. However, a full assessment must explore the possible effect of non-cyclic stresses present as, say, welding residual stresses or as 'system' stresses due to attachment to some other structure. Further, the particular operational environment can markedly influence the response of a material to fatigue loading, this applies to such features as the chemical activity of the fluid, the temperature of operation and nuclear irradiation.

As far as fatigue extension is concerned, the assessment is the same whether it relates to analysis of a detected flaw or to a flaw postulated for design analysis. Therefore we will concentrate on the former situation, i.e. the practical problem of calculating the amount of growth of a flaw of size a_0 detected by some NDT technique.

2. BRIEF DESCRIPTION OF FATIGUE CRACK GROWTH MECHANISMS

Until the 1960s the bulk of established data on fatigue cracking related to initiation, i. e. the formation of a crack from a plain surface subject to cyclic deformation. The surface was regarded as smooth on the microstructural scale and crack formation occurred by microscopic processes. The dividing line between the uncracked and cracked state was a subjective one. Actual formation of a crack is governed by shear stresses local to the surface and occurs where deformation bands intersect the surface producing irregularities some of which develop into intrusions and cracks.

Alternatively the shear may interact with second phase or inclusion particles at the surface with the same effect. The detailed metallurgical processes leading to crack initiation are not relevant to the present review and reference should be made to recent text books (e.g. reference 1).

Fatigue properties were generally measured in terms of $S-N$ curves which related the number of cycles for test specimen failure, N, to the applied fatigue stress amplitude, S. A typical $S-N$ curve is shown in Fig. 1. Its most

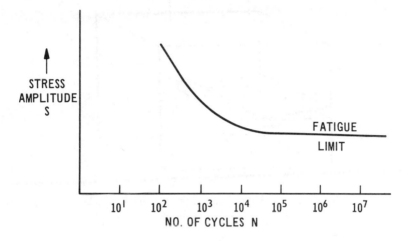

FIG. 1. An idealised $S-N$ fatigue curve. The fatigue stress amplitude S is plotted on a linear scale and the number of cycles to failure, N, on a log scale.

important feature is that of a 'limiting stress or fatigue limit' beneath which failure does not occur. This 'limiting stress' occurs commonly in steels but the 'limit' may be in fact a slowly decreasing stress level. Therefore it is usual to measure the fatigue limit at a given number of cycles, e.g. at 10^7 or 10^8 cycles.

Typically a failure point at a given stress on an $S-N$ curve is the number of cycles when the test specimen breaks in two. Thus N is the sum of the number of cycles for crack 'initiation' and the number of cycles to grow this crack through the section. The relative number of cycles of crack growth depends on the definition of initiation, the applied stress amplitude and the surface condition of the test piece, e.g. at high stress most cycles to failure are associated with crack growth but at low stress with initiation. Normally 'crack growth' is taken to refer to the final stage of cracking which is controlled by the stress normal to the crack faces and follows the shear stress controlled initiation and very early growth stages. Macroscopically, therefore, crack growth is at right angles to the maximum principal stress and is subject to the mode I crack opening displacement in fracture mechanics terms.

For simple descriptive purposes consider the model sharp crack in Fig. 2. This is at the surface of a structure which is periodically loaded to a

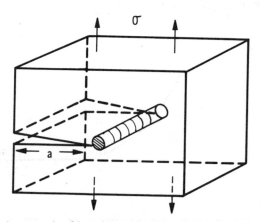

FIG. 2.   A fatigue crack of length a and subject to stress σ. The crack tip plastic zone of radius r_y is shown shaded.

nominally elastic stress level, σ, typically about one half of the material's yield stress. The bulk of the material cycles reversibly in an elastic manner and no fatigue occurs. Local to the flaw tip, however, the applied stress is concentrated and exceeds the yield stress for some distance which depends mainly on the crack depth and the ratio of the applied stress to the material yield stress. The resulting plastic deformation is accommodated in a zone shown schematically and shaded in the figure. On unloading from σ this volume of plastically deforming material decreases in size, the cycle is completed and the crack will have extended by a small increment of length. The mechanism of crack extension is not immediately obvious from a simple picture of reversed plastic flow at the crack tip. Nevertheless there are convincing observations that a crack does extend incrementally during fatigue. These include programmed fatigue tests where the fractographic pattern on the fractured surface exactly reproduced the imposed loading pattern,[2-4] i.e. each cycle created a microscopic striation on the fracture surface and the striation pattern mirrored the loading pattern.

On application of the load, the crack faces in Fig. 2 move apart. This

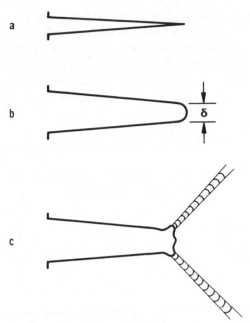

FIG. 3. Diagrammatic representation of the deformation at a fatigue crack tip. In (a) the crack is unloaded. The plastic blunting is shown in (b) and the crack tip opening displacement is δ which also is a measure of the amount of new crack surface. A possible mechanism of striation formed by localised shear bands is indicated in (c).

displacement is partly elastic and partly accommodated by the plasticity at the crack tip. When the plastic zone ahead of the crack is small relative to the size of the ligament then the magnitude of this crack face separation at the crack tip, δ, is given by

$$\delta = \frac{2\pi\sigma_y r_y}{E} \tag{1}$$

where σ_y is the yield stress, r_y the radius of the plastic zone and E is Young's modulus. δ is often referred to as the COD, crack opening displacement. This crack opening causes blunting of the sharp crack tip as sketched in Figs. 3(a) and (b).

Plastic blunting is accompanied by localised shear on bands at an angle to the crack as shown in Fig. 3(c) and this intensely localised shear may account for the striation markings at each cycle and on both crack faces. The new crack surface in Fig. 3(b) produced by plastic stretching during the loading part of the fatigue cycle is thought to produce the increment in crack length for that cycle. On unloading, the crack is resharpened but the plasticity is not completely reversed and all or part of this new surface is recovered for crack growth. The detailed micro-mechanisms during this stage are still uncertain since many contributory events can affect the efficiency of this recovery; for example the adsorption of gaseous atoms on the freshly formed surface will prevent rewelding and enable a large part of the new surface to be retained.

3. QUANTITATIVE CHARACTERISATION OF FATIGUE CRACK GROWTH

3.1. Data Procurement

Fatigue cracks will obviously be most likely to develop at the highest stressed regions of a pressurised component. Penetrations into pressure vessels are prime contenders for concern and serve to illustrate the phenomenon in Fig. 4. This is taken from Miyazono and Shibata[5] who

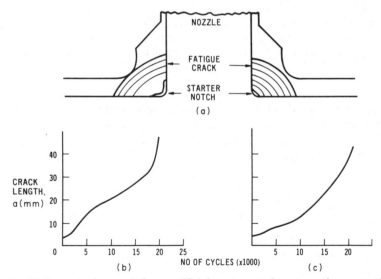

FIG. 4. Fatigue crack growth from artificial starter notches at nozzle penetrations in a pressure vessel. (a) Section through nozzle penetration of vessel showing starter notches and subsequent fatigue cracks. The crack on the left has grown through the wall thickness. (b) Fatigue crack length as a function of number of cycles from left-hand starter notch. (c) Fatigue crack length from right-hand starter notch.

measured the rate at which fatigue cracks grew from small artificial notches at the crotch corner of nozzle penetrations in pressure vessels. The steel vessels were cyclically pressurised using oil until the particular crack shown on the left-hand side of the diagram broke through to the vessel surface. Crack lengths were monitored using an electrical resistance technique and the increase in length of each crack during cycling is shown.

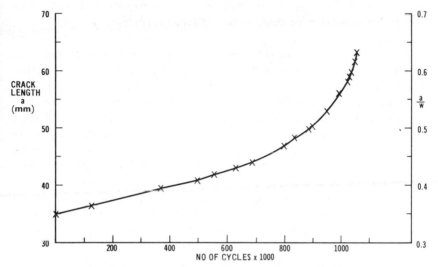

FIG. 5. Fatigue crack growth in a low alloy pressure vessel steel tested at room temperature. The specimen was a 50 mm thick compact tension type tested at constant load with a starting ΔK of 12 MPa m$^{1/2}$ and $R = 0.7$, where
$$R = K_{min}/K_{max}.$$

Laboratory measurements of fatigue growth produce very similar results as demonstrated in Fig. 5. The object of laboratory analyses is to derive this type of fatigue data for the relevant material, characterise it in terms of stress and use it to predict the rate at which cracks such as those in Fig. 4 will grow in service.

Laboratory derivation of fatigue data must simulate operational duty as closely as practicable. Most pressurised components are subjected typically to a large number of cycles throughout their lifetime, the amount of plasticity each cycle being small. Thus we are less concerned in this chapter with 'high strain' fatigue and more with stress controlled fatigue. Recent advances in fatigue experimentation in this latter regime rely on fracture mechanics' techniques since these provide a unified methodology for defect assessments. Thus data procurement borrows fracture mechanics type specimens and data analysis is usually in terms of stress intensity concepts.

There are no UK standard test techniques for fatigue crack growth. BS 3518 on methods of fatigue testing is only concerned with S–N curves (as is the latest British Standard pressure vessel code, BS 5500, issued in 1976). In the US, the ASTM Task Group on Fatigue Crack Growth Rate Testing (from the E24 Committee) is drafting a proposed test method using either compact tension type or centre-cracked tension specimens.

Fatigue test specimens are generally standard fracture toughness compact tension (CTS) or bend specimens which are described in British Standards Institution Draft for Development on Fracture Toughness Testing, DD3. There are standard compliance calibrations for these specimens to allow easy calculation of the stress intensity factor. Crack growth measurements are most conveniently done in servo-hydraulic machines which have a facility for imposing a wide variety of loading cycles automatically. Very high cycle fatigue testing, $> 10^7$ cycles is usually performed on resonance machines. The test specimens are notched by slitting and the slot sharpened by pre-fatigue cracking at low stress levels until the crack is a third to a half of the specimen width. Testing is frequently carried out at constant load so that the stress intensity (stress times square root of crack length) increases as the crack propagates. A useful variation on this is to test at constant stress intensity by decreasing load as the crack grows; it is even possible to use a specially contoured specimen of increasing height designed so that the stress intensity remains constant as the crack grows.

There are several techniques for crack length monitoring during a test. The simplest is by eye and uses a millimetre grid scribed on the specimen edges: visual estimation allows an accuracy of about 0·2 mm. Automatic methods employ (i) an ultrasonic transducer which follows the crack front via a micrometer screw movement, or (ii) a clip gauge across the crack mouth which monitors the increase in specimen compliance as the crack gets longer, or (iii) most commonly, to use an electrical resistivity technique where a current is passed through the specimen and fine leads spot-welded on either side of the crack measure changes in potential as the crack grows—this technique measures changes in crack length to about 0·1 mm.

The raw data are of the form shown in Fig. 5. It is usual to present the data as a log–log plot of growth rate per cycle, da/dN, as a function of stress intensity, K where a is crack length and N the number of cycles. Growth rate is determined simply by taking the tangent at points on Fig. 5 or by computer manipulation of the raw data.

3.2. The Mechanics of Crack Growth

When the data of Fig. 5 are transcribed into terms of growth rate versus stress intensity, Fig. 6 shows the typical result. The fatigue stress amplitude, $\Delta\sigma$, is transformed to stress intensity amplitude ΔK:

$$\Delta K \propto \Delta\sigma = \sigma_{max} - \sigma_{min}$$

where σ_{max} is the maximum stress in the cycle and σ_{min} is the minimum stress

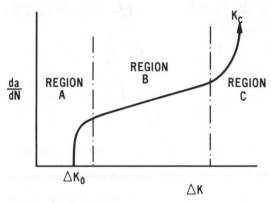

FIG. 6. An idealised log–log plot of fatigue crack growth rate, da/dN, as a function of stress intensity amplitude, ΔK. At very low values of ΔK there is a threshold, ΔK_0, beneath which crack growth cannot be detected and at high values of ΔK crack growth accelerates to unstable fracture at K_c. The curve is divided into three regions A–C.

in the cycle. In practice, for the standard bend and compact tension test pieces, ΔK is calculated as $K_{max} - K_{min}$ and K calculated as in fracture toughness techniques

$$K = \frac{PY}{Bw^{1/2}}$$

where P is the load, B the specimen thickness, w the specimen width and Y the appropriate compliance coefficient $Y = f(a/w)$ as tabulated in DD3.

The typical crack growth curve in Fig. 6 is sigmoidal in shape and may be divided into three parts. Region A is associated with a threshold stress intensity, ΔK_0, beneath which growth is zero or extremely slow and above which growth accelerates sharply. In region B, growth rate is dependent on ΔK raised to a low power, and finally cracking accelerates rapidly at high ΔK values where K_{max} approaches the critical K for failure, K_c. In practice

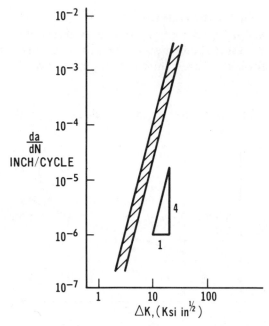

FIG. 7. Fatigue crack growth data for 2024–T3 aluminium alloy from five different sources (after Paris and Erdogan[6]).

these three regions may be more or less obvious and occasionally any one region difficult to define.

Early analysis of data from various sources, mainly on aluminium alloys, drew average lines through the accumulated data which had a slope of about 4,[6] see Fig. 7. Considerable experimental evidence for such a generalised linear relationship accrued and although some very large values for the slope were occasionally reported, a slope of 4 was usually adopted as the norm. The existence of a universal fatigue crack growth law was widely upheld, often known as the Paris Law, and taken as

$$\frac{da}{dN} = C\,\Delta K^4$$

However, careful experimentation and control of the mechanics of the crack growth has demonstrated that the Paris Law describes averaged behaviour and the true form of crack growth relationship is as in Fig. 6. Nevertheless, stage B in Fig. 6 can occur over a wide range of stress intensity amplitude and is described by $(da/dN) = C\,\Delta K^n$ but now n has a value

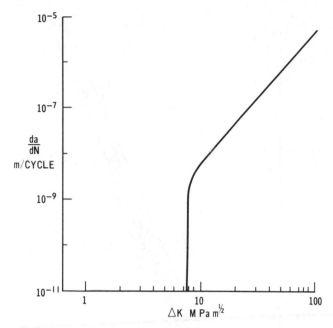

FIG. 8. Fatigue crack growth rate in a low alloy pressure vessel steel as a function of stress intensity factor and at a low value of R. Region A extends from ΔK of 7 to $8\,\mathrm{MPa\,m^{1/2}}$ up to about $10\,\mathrm{MPa\,m^{1/2}}$ and region B from $10\,\mathrm{MPa\,m^{1/2}}$ to just over $100\,\mathrm{MPa\,m^{1/2}}$. It was not possible to test to higher ΔK values and hence obtain region C.

between 2 and 3. For a typical low alloy medium strength pressure vessel steel with yield stress of 400–500 MPa: the threshold value for crack growth, ΔK_0, varies from about 3 to $10\,\mathrm{MPa\,m^{1/2}}$ depending on mean stress (see Section 4.2); stage B extends from about 10 to approximately $100\,\mathrm{MPa\,m^{1/2}}$ and can be described by $(da/dN) = C\,\Delta K^n$ where $C \sim 3 \times 10^{-11}$ (where the growth rate is measured in units of m/cycle and ΔK in $\mathrm{MPa\,m^{1/2}}$); and $n \sim 2.5$. An actual curve for a low alloy pressure vessel steel is reproduced in Fig. 8.

In general there is rarely adequate data available for any specific material to make a full analysis of crack growth, so a pessimistic† assessment is often made. This ignores the threshold and assumes a single power law exists between stress intensity and growth rate from very low stress intensities up

† A 'pessimistic' assessment overestimates the amount of crack growth and hence errs on the safe side.

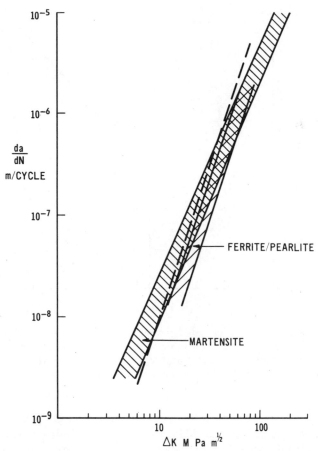

FIG. 9. Fatigue crack growth rate data band for high strength martensitic and lower strength ferrite/pearlite steels (after Barsom *et al.*[7,8]). Superimposed is the upper bound line for a wide range of low alloy steels and weldments (after Richards and Lindley[9]).

to values of ΔK about 75 % of K_c. There are not large variations in fatigue crack growth rates within given classes of materials; this statement refers to the large range of crack growth in region B of Fig. 6. For many purposes fatigue crack growth in steels or aluminium alloys in this region is little affected by yield strength or by their state of heat treatment. Therefore generalised crack growth relationships can often be used keeping in mind particular guidelines and exceptions touched on in Section 4. For example, data for crack growth in air at room temperature are plotted in Fig, 9 for a wide range of constructional steels. Two bands enclose the data points of Barsom and coworkers for a series of ferrite/pearlite steels of proof strengths from 250 to 400 MPa[7] and a series of high strength martensitic steels of proof strengths from 600 to 1300 MPa.[8] These bands are very narrow and overlap. Also shown is the upper bound to the data collected by Richards and Lindley[9] for a wide range of steels including weldments with proof strengths from 200 to 900 MPa; these data are a very good agreement with the rest. The upper bonds to the growth data are given by the following expressions:

$$\frac{da}{dN} = 9 \times 10^{-12} \Delta K^3 \text{ for the ferrite/pearlite steels}$$

$$\frac{da}{dN} = 1 \cdot 7 \times 10^{-11} \Delta K^{2 \cdot 25} \text{ for the martensitic steels}$$

$$\frac{da}{dN} = 1 \times 10^{-11} \Delta K^3 \text{ for Richards and Lindley data}$$

where growth rate is in m/cycle and ΔK in MPa m$^{1/2}$. These are very close and if data for specific steels were lacking then the general expression $(da/dN) = 10^{-11} \Delta K^3$ would be a good approximation.

There is theoretical support for such crack growth dependencies. Within the above data bands, the actual experimental results for tough ferritic constructional steels follow a power dependence of 2–2½. Fatigue crack growth in the higher strength, more brittle steels can be enhanced by crack growth mechanisms other than striation formation and hence the power dependence is nearer 3. In Section 2 it was shown how fatigue crack growth was controlled by the crack tip opening displacement described by eqn. (1). The plastic zone size. r_y in this expression can be rewritten in terms of the alternating stress or stress intensity so that the displacement, and hence the crack growth rate, is proportional to ΔK^2.

4. ANALYTICAL ASSESSMENT OF FATIGUE CRACK GROWTH

4.1. Generalised Analyses
Fatigue crack growth assessments start from the simple mechanics crack growth 'law'

$$\frac{da}{dN} = C \Delta K^n \tag{2}$$

The complexity of any analysis depends on the importance of the assessment, ranging from a 'back of envelope' check to a sophisticated computerised examination of the safety of a nuclear reactor pressure vessel over its 40 year life. Nevertheless in both cases the analyses use eqn. (2). For consistency in the various examples quoted we will use the general expression given in Section 3 for steels, i.e. $C = 10^{-11}$ and $n = 3$ (growth rate in m/c, ΔK in $MPa\,m^{1/2}$).

At the back of envelope stage, consider a particular crack subject to a pressure loading which increases the stress intensity at the tip of the defect from zero to $10\,MPa\,m^{1/2}$. In this case the amount of crack growth da/dN in the cycle is $10^{-11} \times 10^3\,m = 10^{-5}\,mm$. For comparison, a much larger stress intensity amplitude ΔK of $100\,MPa\,m^{1/2}$ would produce crack growth of $10^{-11} \times 100^3\,m$, i.e. $10^{-2}\,mm$.

A simple calculation like the one above provides an indication of the instantaneous amount of cracking to be expected for a particular loading cycle. However, it is often necessary to sum the extent of crack growth resulting from the loading cycles expected over the life of the component in question. This requires a simple (but often tedious) iteration of the above calculation. A useful formalism of the procedure is given in the *American Society of Mechanical Engineers (ASME) Boiler and Pressure Vessel Code*, Section XI 'Rules for Inservice Inspection of Nuclear Power Plant Components' (1974 Edition) (see also Section 4.4). Non-mandatory Appendix A to Section XI defines a procedure for evaluation of a defect found in a thick section ferritic steel pressure vessel. This requires the assessor to consider every transient condition which will impose a cyclic pressure or thermal stress on the vessel as a result of normal operation, testing or upset conditions.† The stress amplitude at the defect must now be calculated for each transient condition, e.g. start up–shut down or power increase/decrease. Using eqn. (2) the amount of crack growth, Δa, is calculated for the first transient. Δa is now added to the initial crack length a and ΔK recalculated for the new but geometrically similar crack length $a + \Delta a$. The increment in crack length is now calculated for this slightly longer crack length as a result of the next transient. Thus the crack length is progressively updated for each and every transient in approximately chronological order. The final crack length is the end of life flaw size, a_f, which has to be compared with the critical flaw size for fast fracture. A

---

† Upset conditions are abnormal operational transients like reactor trips which occur a predictable number of times and hence can be designed against (as opposed to emergency or fault conditions).

TABLE 1

Cycle Number	Initial crack length (mm)	$\Delta\sigma$ (MPa)	ΔK (MPa m$^{1/2}$)	Δa (mm)	New crack length (mm)
1	25·00	250	72·06	0·0037	25·004
2	25·004	175	50·45	0·0013	25·005
3	25·005	175	50·45	0·0013	25·006
4	25·006	250	72·07	0·0037	25·01
5	25·01	175	50·46	0·0013	25·011
6	25·011	175	50·46	0·0013	25·012
7	25·012	175	50·46	0·0013	25·014
·	i.e. 50 cycles at $\Delta\sigma = 175$				
53	25·072	175	50·52	0·0013	25·073
54	25·073	175	50·52	0·0013	25·074
55	25·074	250	72·17	0·0037	25·078

Crack growth of an assumed longitudinal surface flaw of aspect ratio $(a/l) = 0.2$ in the 100 mm thick wall of a pressure vessel. The flaw is subject to mode I stresses in the hoop direction. K is calculated from ASME XI Article A3000; the membrane stress σ_m need only be considered, i.e.

$$K = \sigma_m M_m \left(\frac{\pi a}{Q}\right)^{1/2}$$

where a = flaw depth; l = flaw length; t = wall thickness (confusingly, ASME XI uses t for wall thickness but elsewhere in this chapter this dimension and test specimen width is w, as is more usual); M_m = correction factor for surface flaws (Fig. A-3300-3)—for $(a/l) = 0.2$, and $(a/t) = 0.25$, $M_m = 1.15$; and Q = flaw shape parameter (Fig. A-3300-1)—for $(a/l) = 0.2$, $Q = 1.25$. The crack shape is assumed to remain constant at $(a/l) = 0.2$ as the crack grows by fatigue.

simple example of an iterative crack growth calculation is given in Table 1.

A detailed application of the above cumulative analysis was published by Riccardella and Mager[10] who assessed the fatigue life of PWR vessels. They spread the total number of expected loading cycles evenly over the 40 year life of the vessel; this pattern was then divided into forty blocks of one-year each so that each annual schedule contained a proportion of all the transients (except those cycles which are only applied in the pre-operational commissioning period). Roughly, the design transients over the 40 year life comprised 10^6 low stress cycles due to small but frequent temperature changes and about 3000 high stress operational cycles. Notional cracks were considered to be present at four highly stressed design details of the pressure vessel and the cyclic stress amplitude was calculated for these sites and each of the design transients. Although the likely crack shape is 'thumbnail', it was approximated conservatively by a continuous surface crack and stress intensity factors were calculated using the compliance functions for single edge notch specimens which deform by combined tension and bending. Fatigue crack growth data in the form of eqn. (2) were obtained using the relevant pressure vessel steels immersed in a simulated reactor water at the operating temperature. These data were applied to the postulated cracks up to 1 in long at the four critical sites in the iterative manner described in the previous paragraph and the total amount of crack growth over 40 years tabulated. The tedious iterative calculations were performed by computer. The maximum amount of crack growth occurred at the highly stressed intersection between the cooling water nozzle and the vessel wall; after 40 years a 0·50 in (12·7 mm) long crack was estimated to grow to 0·53 in (13·5 mm) in length and a 1·0 in (25·4 mm) crack to 1·37 in (34·8 mm). This amount of growth was considered to be nearly negligible.

Within such a large complex fatigue analysis there are many uncertainties and approximations; some are considered below. (i) Stress analysis is expensive and usually approximate; stress intensity factor calculations are also approximate but the modelling of semi-elliptical cracks in a cylindrical geometry by continuous edge cracks in a plate is conservative. (ii) The number and magnitude of transients actually experienced by plant during operation may differ from those predicted by the designers. (iii) If a crack is found by inspection its size may be established accurately enough; however, for a design assessment, like the above example, the maximum size of crack which inspection can *guarantee* to find in a thick section weldment may be difficult to define (and > 1 in). (iv) Finally, the crack growth data may not represent upper bound values. This will be discussed in detail later but the 'mean' stress in the fatigue cycle can influence crack growth in an aggresive environment although it has little effect in air. The mean stress is measured by the ratio of the minimum to the maximum stress in the cycle and is described by R:

$$R = \frac{\sigma_{min}}{\sigma_{max}} = \frac{K_{min}}{K_{max}}$$

Thus a cycle from, say, -10 to 10 is described by $R = -1$; from 0 to 20 by $R = 0$; from 80 to 100 by $R = 0·8$, etc. For the above example it is now known that if account were taken of mean stress–environmental effects and realistic cycle frequency then faster crack growth rates would result. For example, the crack growth data used in reference 10 were obtained at a test frequency of 1 Hz and a later analysis[11] repeated the assessment using faster crack growth data obtained at 0·02 Hz but at a low value of R. Whereas a 0·50 in (12·7 mm) crack at the nozzle–vessel wall intersection had been predicted to grow to 0·53 in (13·5 mm) in length after 40 years, the faster growth data predicted a final crack size of 0·70 in (17·8 mm).

Because of the nuclear safety connotations of this work, these complex fatigue calculations were repeated by the United Kingdom Atomic Energy Authority.[12] The same analytical route was used, similar (but not identical) operational transients were assumed and the same crack locations assessed. The major difference was in the fatigue crack growth data; although the low frequency (0·02 Hz) data were used the data were modified by inclusion of results obtained using higher values of R which produce a faster growth rate in the aqueous environment. However, in this analysis the more relevant semi-elliptic crack geometry was assumed. Despite this, significant cracking resulted, e.g. a 0·50 in (12·7 mm) crack was predicted to grow to 0·82 in (20·8 mm) and a 1·0 in (25·4 mm) crack to 3·0 in (76·2 mm); this amount of crack growth is no longer 'nearly negligible'.

Although the iterative process for calculating crack growth is tedious, especially if there is not access to a computer, it has the advantage that it accurately accounts for the change in compliance of the structure as the crack increases in length. It is simple to integrate eqn. (2) over the required number of cycles and obtain the total crack growth directly. However, the integration can only be done simply if the compliance does not vary significantly over the range of crack length involved.

Equation (2) relates crack growth to stress intensity amplitude

$$\frac{\mathrm{d}a}{\mathrm{d}N} = C \, \Delta K^n$$

In the general case, $K = \sigma a^{1/2} Y(a/w)$ where $Y(a/w)$ is the compliance function for the particular geometry, a is crack length and w is the width. However, for the simple case of a crack embedded in an infinite body, $Y(a/w) = \pi^{1/2}$; thus substituting for ΔK in eqn. (2) gives

$$\frac{\mathrm{d}a}{\mathrm{d}N} = C(\Delta\sigma\pi^{1/2}a^{1/2})^n = C\pi^{n/2}\,\Delta\sigma^n a^{n/2}$$

Integrating between the initial crack size, a_0, and a final crack size, a_f, gives

$$\int_{a_0}^{a_f} a^{-n/2}\,\mathrm{d}a = C\pi^{n/2}\,\Delta\sigma^n \int_{N=0}^{N=N_f}\mathrm{d}N$$

Therefore

$$\frac{2}{2 - n} [a_f^{1 - n/2} - a_0^{1 - n/2}] = C\pi^{n/2} \Delta\sigma^n N_f$$

This expression cannot be used for $n = 2$, but for $n = 3$:

$$(a_0^{-1/2} + a_f^{-1/2}) = \frac{C\pi^{3/2}}{2} \Delta\sigma^3 N_f \qquad (3)$$

(and when $C = 10^{-11}$) $= 2 \cdot 8 \times 10^{-11} \Delta\sigma^3 N_f$

The use of the iterative technique was illustrated in Table 1 and we can now compare these results with the approximate integration described above. The table shows the amount of fatigue crack growth for the simplified case of a flaw in a pressure vessel subject to proof pressurisations (to a stress of 250 MPa) and a series of repeated operational cycles to a stress of 175 MPa. The extent of crack growth is very small and the corresponding change in the geometrical parameters in the calculation of ΔK insignificantly small. Accordingly if eqn. (3) is used to calculate the amount of growth from cycle 5 to 54, i.e. 50 cycles at 175 MPa, close agreement is obtained because of the insignificant change in $Y(a/w)$, i.e. from Table 1 the crack grows from 25·01 to 25·074 mm using the iterative technique and using eqn. (3) from $a_0 = 25·01$ to $a_f = 25·070$ mm.

Equation (3) may be used in several ways. In the above calculation the integration was in the forward direction, i.e. from a_0 to a_f for a given number of cycles. It may also be used to calculate the number of cycles, N_f, which would cause an initial known defect, a_0, to grow to the critical size for fast fracture, a_f, as calculated from fracture toughness data. Further, by inverting the calculation, the size of initial defect, a_0, which has to be detected by non-destructive inspection prior to plant commissioning, can be calculated knowing the critical size for fast fracture, a_f, at end of plant life, N_f.

The amount of fatigue crack growth in the above example is insignificant and if either ΔK or N were larger then the change in crack length could have become significant with corresponding errors in the integration route. In complex and lengthy analyses, however, it is often possible to break down the total cyclic life into blocks of transients so that the amount of growth per block can be integrated and the total growth obtained by iteration. The actual computation involved depends on the sophistication of the analysis and hence may be performed on simple or programmable pocket calculators or require the largest computers. For example, in the author's laboratory a sophisticated program called FATPAC has been developed[13] based on eqn. (2). In this program, if the number of cycles, N, is less than 100 then an iterative calculation is performed, if $N > 100$ then growth is integrated over $N/100$ pieces. The crack growth 'law' of Fig. 6 is used as input to the computer with a combination of up to five linear da/dN versus ΔK ranges, i.e. stages A, B, C or more. The program can accept up to 80 types of cycle in a block of loadings and up to 40 blocks in the total life to be assessed. The strength of this particular program is its ability to call up a

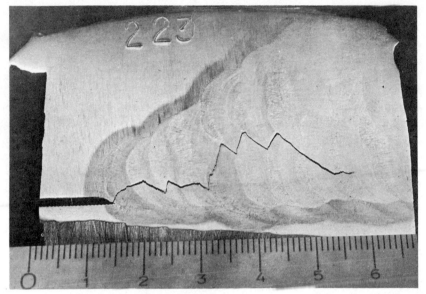

FIG. 10, Cracking extending from the unfused land in a boiler drum/nozzle weld (reference 43).

program called FRACPAC as a subroutine which can calculate the stress intensity and hence ΔK for a wide variety of geometries and for any stress field.

An actual example from the power generation industry illustrates simply the use of eqn. (3). Some large 150 mm thick boiler drums which contain steam at 18·3 MPa (2650 psi) and 350 °C were found in 1968 to be badly cracked, see Fig. 10. The cracks had formed during stress relieving from an unfused land at a butt weld connecting the nozzles to the shell. Fracture mechanics showed that there was little chance of sudden unstable failure and that the fatigue crack growth associated with a small number of pressure cycles was acceptable. Enforced shut down of all the defective plant was thus avoided and repairs were made in a phased manner. Considering further the same example, an inspection criterion for the repair welds was developed by deriving the critical crack size for fast fracture and making an allowance for fatigue crack growth. The critical crack size for the nozzle geometry was conservatively estimated to be 12·0 mm using a fracture toughness of 110 MPa m$^{1/2}$ for the mild steel repair weld metal. The remaining service requirement for the assessment fatigue life of the vessels was defined as comprising 10^4 hot starts (i.e. pressurisation from 12·4 to 18·3 MPa) interspersed with 300 pressurisations from cold (i.e. 0 to 18·3 MPa). Because of uncertainties in the analyses at the time, the stresses were pessimistically taken as $\Delta\sigma = 51$ and 160 MPa for the two types of cycle. Then from eqn. (3), a_0 is 11·8 mm i.e. a starting flaw 11·8 mm in size would grow to 12·0 mm after 10^4 hot and 300 cold starts. Thus the minimum size of defect which NDT had to be able to guarantee detection after weld repair was 11·8 mm. In this instance the allowance for fatigue (0·2 mm) is insignificant in terms of inspection. As a rule the higher the fracture toughness of the material in question, then the larger is the difference between a_f and a_0 (because a_f, and hence ΔK, are relatively large).

The example above, like the others, assumes that fatigue crack growth is cumulative, i.e. that there is no interaction between different types of cycle. This assumption is not generally true since a large cycle will limit the amount of growth from immediately following, smaller cycles. For instance, repeat over-pressure tests on a pressure vessel will retard fatigue crack growth from several subsequent normal duty cycles. However, in practice where thick section components are concerned the amount of retardation will be very small at normal over-pressure stresses. This is illustrated in Fig. 11 using 25 mm thick specimens made from a low alloy pressure vessel steel. Cracks were propagated throughout the test at a ΔK of 30 MPa m$^{1/2}$ and $R = 0.05$ and subjected to single overload cycles of 50, 60 then 70 MPa m$^{1/2}$. Only a small retardation occurred at 50 MPa m$^{1/2}$ but the extent of retardation increased markedly with the overload stress intensity and lasted for several thousand cycles. During the retardations, crack growth continued but at a very slow rate. Overload effects are generally ignored in pressure vessel fatigue analyses; this should aid the pessimism of the final result. The effect of a periodic over-pressure test will be small but in a complex loading pattern, as in the PWR vessel example, a large pressure transient could significantly slow down the rate of crack growth resulting from subsequent minor pressure fluctuations. This topic is of importance in the aero industry, however, where the normal loading spectrum contains large cycles

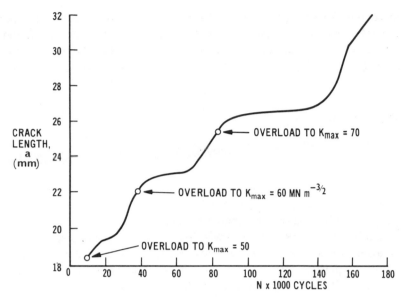

FIG. 11. Fatigue crack growth retardation due to periodic overload cycles. 25 mm thick compact tension specimens of low alloy pressure vessel steel tested at room temperature (after Bernard, Lindley and Richards[44]).

interspersed with much smaller cycles. Much of the current research is directed at relatively thin section aluminium alloy but spin off from this rescarch is likely to aid understanding of fatigue processes in general.

The reasons for overload or spectrum loading effects are not yet clear. Qualitatively an overload will produce an increase in the crack tip plastic zone size, r_y, since

$$r_y = \text{const.} \left(\frac{K}{\sigma_y}\right)^2 \qquad (4)$$

On unloading from the overload, the surrounding elastic material will force the yielded zone into compression. The resulting residual compressive stresses will limit the amount of crack tip opening and hence crack growth during subsequent smaller cycles. The retardation should be a function of the ratio of the overload and reload plastic zone sizes from eqn. (4) or, equivalently, the ratio of crack tip openings from eqn. (1).

An alternative, but not unrelated, explanation stems from observations made by Elber[14] on crack face closure effects. During fatigue crack propagation in thin aluminium plates, the crack faces appeared to remain in contact and did not open until part way up the loading cycle (even though the whole cycle was in tension). Thus crack growth should be governed by an effective stress intensity ranging from the stress intensity at which crack opening occurs, K_{op}, to the maximum stress intensity, i.e.

$$\Delta K_{eff} = K_{max} - K_{op} \tag{5}$$

rather than $\Delta K = K_{max} - K_{min}$. Thus

$$\frac{da}{dN} = \text{const.} \ (\Delta K_{eff})^n$$

Overload retardation follows when K_{max} of the smaller cycles is less than K_{op}. Some success appears to result from the use of crack closure analyses especially in the spectrum loading field (see e.g. references 15–17). Physically, crack closure is thought to result from local residual displacements at the crack surfaces produced when the crack grows through its plastic zone. Experimental measurement of K_{op} is difficult in practice and it is not clear how relevant it is to thick section structures where normally any deformation local to a crack is in plane strain. For instance, measurements on steel specimens of varying thickness[18] showed that as long as the fatigue cracks were growing in plane strain then no crack closure occurred unless the load was decreased beneath the minimum load level of the cycle. Nevertheless when plane stress deformation became significant then crack closure was detected above K_{min}. An up to date, brief review of crack closure has been published by Bachmann and Munz.[19]

4.2. Threshold Effects

In the examples so far it has been assumed that crack growth can occur at very low values of ΔK, i.e. eqn. (2) applies down to $\Delta K = 0$. However, it was shown in Fig. 6 that crack growth becomes zero or extremely slow beneath a threshold stress intensity ΔK_0. Although there is now a good data bank for ΔK_0, the understanding of threshold effects is lacking.

There are some similarities between ΔK_0 and the conventional fatigue limit. For example, it is not clear whether growth is non-existent beneath ΔK_0 or just too low to be measured. Therefore ΔK_0 is usually measured as the stress intensity beneath which there is no detectable crack growth after a given number of cycles; this is usually about 10^7 although measurements have been made to over 10^8 cycles. Experimental values of ΔK_0 for various materials have been collected in the literature, see e.g. references 20 and 21; the most notable feature, especially for steels, is the marked decrease in ΔK_0

as the mean stress R is raised, see Fig. 12.[22] This diagram shows a compilation of data from tests on mild steels and low alloy steels including pressure vessel steels. Lower thresholds do exist, for instance ΔK_0 for high strength martensitic steels can fall well below the lower bound of Fig. 12, even at low R values (22). Data are much more sparse for negative values of R

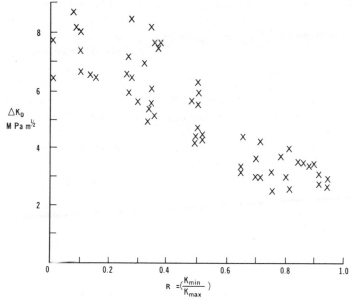

FIG. 12. The threshold stress intensity for fatigue crack growth, ΔK_0, plotted as a function of $R \, (= K_{min}/K_{max})$, after reference 22.

($K_{min} < 0$) but the indications are that the thresholds are increased as R goes from 0 to -1.† There is a shortage of data on fatigue crack growth thresholds for weldments which is surprising when one considers that most defects in structures are associated with welds. Qualitatively one would anticipate that refined welds would have similar thresholds to those of Fig. 12 but that coarse grained, hard, weld microstructures could have low values of ΔK_0.

In practice threshold data are not widely applied to pressurised structures like pressure vessels; most assessments assume eqn. (2) applies to very low

† When $K_{min} < 0$, then ΔK is strictly $K_{max} - K_{min}$ but in practice it is likely that fatigue cracking is governed by $\Delta K_{eff} \sim 0$ to K_{max}.

ΔK values. This is a safe assumption especially in the absence of reliable data. It is more important to take account of threshold effects for structures where fatigue crack growth is fast and the life-limiting criterion is whether or not cracking will initiate from a pre-existing defect. A good example of the application of this approach to rotating plant has been published in reference 23.

Normally ΔK_0 is measured experimentally on fracture mechanics type specimens containing relatively large cracks, i.e. $a/w \sim 0.5$. However, in practice one may have to analyse situations where the crack is very small relative to structural dimensions. The localised plastic deformation at the near surface crack tip may therefore be different from that at a deeply embedded crack tip with correspondingly different threshold behaviour. Indeed there is some experimental evidence that ΔK_0 measured at very small crack depths (0·025–0·25 mm) is less than at larger depths (0·5–5·0 mm).[24]

Normally, once the stress intensity at a crack exceeds ΔK_0, that crack will continue to grow in the structure. However, the above evidence for a crack size effect on ΔK_0 would give conditions whereby a very small crack could arrest as ΔK_0 exceeds ΔK. Similarly if a fatigue crack initiated at a site of stress concentration and grew into a rapidly decreasing stress gradient then a non-propagating crack could also result. Quantitatively this can be assessed from a stress analysis of the particular design detail and thence comparing a plot of ΔK versus crack length with the value of ΔK_0 for the relevant conditions.

The manner in which intentional or unintentional stress concentrations like notches influence conventional fatigue strength and crack initiation has been well reviewed in Frost, Marsh and Pook's textbook.[1] However, fatigue crack growth close to stress concentrations is much less well documented. Once the fatigue crack grows beyond the zone where the stress concentrator locally perturbs the general stress field, e.g. for a hole beyond about one radius, then one can assume the crack has an effective length equal to the hole plus the crack length and is subjected to the general stress field. Broek has demonstrated that this is a reasonable assumption.[25] When fatigue cracks are within the local stress field of a notch, however, then due account must be taken of the complex stress field. Smith and Miller[26] have recently extended the above ideas and developed an analysis with experimental validation for the near notch field situation. This is based on the hypothesis that two fatigue cracks, one of length l and growing from a notch and the other length L and growing from a plain surface under an identical general stress, are subject to the same controlling conditions when their velocities are the same. Thus the notch contribution, e, to the crack of length l is simply

$$e = L - l$$

Since the two cracks are subject to the same general stress, this notch factor can be expressed in terms of stress intensities. This leads Smith and Miller to a generalised expression for the contribution to a crack length l from a notch of depth D and root radius ρ:

$$e = 7.691 \left(\frac{D}{\rho} \right)^{1/2}$$

For cracks sufficiently large to be beyond the local stress field of the notch then the notch contribution e becomes equal to D, the notch depth.

4.3. The Final Stages of Crack Growth

Equation (3) related the number of cycles to failure, N_f, to the initial crack size, a_0, and the critical crack size for unstable fracture, a_f. In this expression, N_f is sensitive to a_0 and less so to the final crack size, a_f, which is governed by the material fracture toughness. The fatigue life integration, however, assumed that the growth exponent, n, remained constant whereas it was seen in Fig. 6 that growth accelerated rapidly to failure as K_{max} approaches the critical stress intensity for fast fracture. This acceleration in crack growth is the consequence of additional fracture processes like void formation, cleavage or intergranular cracking, becoming active at these higher stress intensity levels.[9] Clearly fatigue crack growth in general is not only a function of ΔK, but of both ΔK and K_{max}.

The simplest way to account for the final stage of fatigue crack growth is to impose an upper limit for K_{max} of about 75% of the critical K. This results in conservative analysis and should ensure the absence of potentially unstable failure modes. Several empirical expressions have been proposed to account for the accelerating stage; these usually involve the inclusion of terms in eqn. (2) which ensure that the growth rate becomes infinitely fast as K_{max} approaches the critical stress intensity. For instance Forman et al.[27] proposed

$$\frac{da}{dN} = \frac{C \Delta K^n}{(1 - R)K_{Ic} - K_{max}}$$

and Heald et al.[28] suggested a similar form

$$\frac{da}{dN} = \text{const.} \left[\frac{\Delta K^4}{\sigma_1^2 (K_{Ic}^2 - K_{max}^2)} \right]^n$$

where σ_1 is a flow stress $\sim (\sigma_y + \sigma_u)/2$ (σ_y = yield strength, σ_u = ultimate tensile strength) and $n \sim \frac{3}{4}$.

Finally, for completeness some workers have tried to produce an expression which will describe the complete crack growth curve from threshold to final acceleration. This is usually done by replacing ΔK by an effective value ($\Delta K - \Delta K_0$): thus Irving and McCartney[29] have recently suggested

$$\frac{da}{dN} = \text{const.} \left[\frac{\Delta K^2 (\Delta K^2 - \Delta K_0^2)}{\sigma_y^2 (K_{Ic}^2 - K_{max}^2)} \right]$$

4.4. Environmental Effects on Fatigue Crack Growth

The influence of environment is normally taken to refer to the effect of atmosphere or fluid in contact with the material in question. There have been two recent meetings devoted to environmental fatigue, one in 1971[30] and one in 1977.[31]

Where the contacting environment detrimentally affects fatigue life, the phenomenon is often termed corrosion fatigue. Early experimentation in corrosion fatigue concerned environmental effects on fatigue 'life', i.e. initiation plus propagation. In general it was believed that environmental effects were due to the creation of cracks at lower stress levels or at shorter times than in a benign environment and that subsequent propagation of such cracks was relatively little affected. Within the last decade, however, much research has been stimulated by aerospace, submarine, nuclear or energy interests, for instance, and has revealed that crack growth can be significantly accelerated by environments (some of which like pure water hardly qualifying as 'aggressive'). Normally the crack growth threshold ΔK_0 is little affected by environment.

Where a structure must be in contact with a strongly corrosive environment then particular attention is paid to material selection or protection. Most fatigue crack growth research, however, has been concerned with mildly aggressive environments like dilute aqueous solutions or gases. As yet there is no generally applicable and quantitative theory for environmental fatigue. The somewhat artificial distinction is often made between fatigue crack growth at K_{max} values above and beneath K_{ISCC}, the stress intensity level beneath which no observable stress corrosion cracking under static loading is observed. Above K_{ISCC} environmental fatigue crack growth is then taken to be the superposition of 'inert' or mechanical fatigue and static stress corrosion. Beneath K_{ISCC} a superposition model cannot apply. This latter situation usually applies to pressurised components like

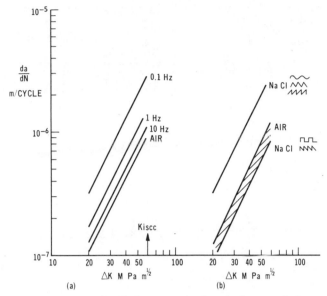

FIG. 13. The influence of cycle frequency and wave shape on fatigue crack growth rate in maraging steel, after Barsom.[32,33] In (a) sine wave shape was used. In (b) all the data were obtained at 0·1 Hz and the appropriate wave shapes are shown.

pressure vessels or pipelines where K_{ISCC} in the operating environment is high (except where high strength materials are used). Any increase in crack growth rate is proportional to the fatigue life, i.e. an acceleration by a factor of 5 in crack growth rate reduces the time for failure by a factor of 5.

It is now known that the most important features influencing environmental effects are cycle frequency and stress ratio. More precisely it appears to be the strain rate during the loading part of the cycle that is important rather than cycle frequency *per se*. For example, Fig. 13 summarises some of the important work of Barsom[32,33] on high strength 12% Ni maraging steel (yield strength 1270 MPa) tested in aerated 3% sodium chloride aqueous solution. In Fig. 13(a) (after reference 32) we see that crack growth rates are little affected by the environment at high frequencies but as the frequency decreases to 10^{-1} Hz there is a factor of three acceleration in growth rate. This effect reflects the time dependency of the electrochemical processes at the crack tip when the cyclic plasticity ruptures the oxide film and allows dissolution and/or hydrogen ingress and transport. Figure 13(b) (after Barsom[33]) gives insight to this frequency effect: the lower narrow scatter band contains the air data for the same

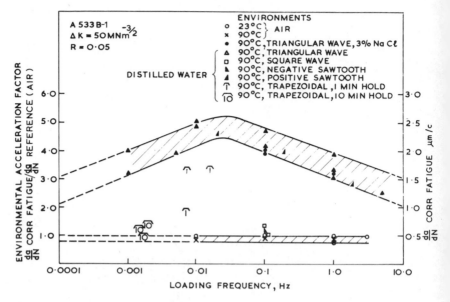

FIG. 14. The influence of cycle frequency, wave shape and hold time at peak stress on fatigue crack growth rate in low alloy pressure vessel steel (after reference 34).

12 % Ni steel plus growth rates in 3 % sodium chloride at 10^{-1} Hz but using square and negative saw-tooth wave shapes. The faster growth curve represents data from tests with sinusoidal (as in (a)), triangular and positive saw-tooth wave shapes. Thus the square and negative saw-tooth waves with very fast ramp loading rates are equivalent to the fast cycling frequency ($\sim$ 10 Hz).† These effects have been confirmed for a much lower strength pressure vessel steel, A533B Class 1, by Atkinson and Lindley.[34] Their results in aerated distilled water and sodium chloride (which gave similar results) are reproduced in Fig. 14. This work shows two additional features. Firstly, the frequency effect passes through a maximum at about 10^{-2} to 10^{-1} Hz; secondly, although the influence of ramp loading rate is as before, if a hold period is introduced at the peak stress intensity of the cycle then the environmental acceleration is progressively diminished.

In practice, therefore, assessment of pressure vessel fatigue crack growth

† The time for one triangular wave equalled that for one sinusoidal wave, and the ramp loading rate for the positive saw-tooth and the unloading rate for the negative saw-tooth cycles equalled the rates for the equivalent parts of the triangular wave.

must take account of the transient loading rate, i.e. strain rate, rather than loading frequency *per se*. Further, we now have another degree of safety built in to such an assessment since pressurised components are not usually subjected to continuous cycling but have hold periods after the loading transient. The high strength and low strength steels described above show very similar environmental response even though the detailed cracking mechanisms are thought to differ. Nevertheless this effect should not be considered to have universal applicability. For instance, when titanium alloys are tested in similar environments and frequency ranges, there is little or no effect of frequency on fatigue crack rate at low ΔK values and only above a particular ΔK value does fatigue crack growth rate increase with decreasing frequency.[35,36]

The concepts of environmentally accelerated fatigue crack growth were introduced in Section 4.1 when discussing the assessment of the PWR primary pressure vessel. The large amount of data collected for this analysis has led to inclusion of fatigue crack growth laws in non-mandatory Appendix A of Section XI of the *ASME Boiler and Pressure Vessel Code* which deals with methods of analysing the acceptability of defects found during inspection. The fatigue crack growth curves are reproduced in Fig. 15 and are described as upper bound curves intended to be very conservative. They refer to low alloy pressure vessel steel of the ASTM A533 and A508 type (plus weldments) of yield strength 345 MPa or less. The growth laws are given as

$$\frac{da}{dN} = 0.0267 \times 10^{-3} \Delta K^{3.726}$$
for air environments

$$\frac{da}{dN} \text{ in microinch/cycle}$$

and

$$\frac{da}{dN} = 0.3795 \times 10^{-3} \Delta K^{3.726}$$
for water environment

$$\Delta K \text{ in ksi in}^{1/2}$$

at low frequency and at about 300 °C. (The very precise value of exponent certainly cannot be justified!) Note that because we still suffer from a maddening lack of uniformity in measuring units, the constants in front of ΔK change to 4.77×10^{-13} for air environment and 6.78×10^{-12} for water environment when growth is in m/c and ΔK in MPa m$^{1/2}$ ($= $ MNm$^{-3/2}$). The crack growth rate in the high temperature water is about $\times 15$ faster than in air.

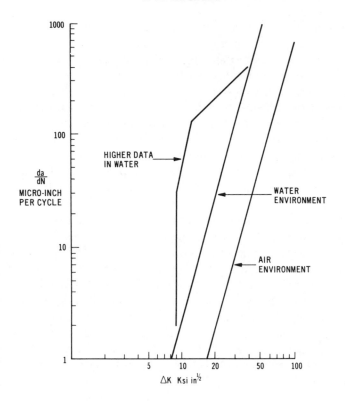

FIG. 15. Upper bound fatigue crack growth data reproduced from ASME XI Appendix A. The air environment line is for sub-surface flaws and the water environment line is for surface flaws. The most recent data obtained at high R values in water (reference 37) are shown for comparison.

Although these ASME fatigue crack growth curves in Fig. 15 are from the 1974 edition of the Code and they were repeated identically in the 1977 edition, they are not conservative. The data were obtained at relatively low R values, usually about 0·1. Occasionally in practice fatigue cycling occurs at higher mean stress values. For instance, where a defect is present at a weld then significant residual stresses may be present and influence the ratio of K_{min} to K_{max}, i.e. R. We have seen previously that R has a marked effect on the threshold for fatigue crack growth but in the absence of environment R has little influence on crack growth rates at ΔK values greater than ΔK_0. The most recent data for reactor pressure vessel steel in high temperature water[37] are at $R = 0·7$ and have been superimposed on Fig. 15; these data markedly exceed the ASME XI water fatigue crack growth line and illustrate the sensitivity of environmental crack growth to R. These new data were obtained at about 10^{-2} Hz and it appears that, as before, this frequency produces the maximum acceleration in environmental growth rate.

The fatigue crack growth data for reactor pressure vessel steel in water were collected at a reactor operating temperature which is just under 300 °C. In general, increasing solution temperature from 20 °C to about 100 °C causes an increase in crack growth rate.[34] However, in pressurised systems the accelerating effect of water appears to peak at about 200 °C whereafter the growth rate decreases up to about 280 °C.[38]

There is some uncertainty at present about the sensitivity of environmental fatigue crack growth to detailed material microstructure. Crack growth in the absence of environmental effects is little affected by microstructure. Evidence from the programme in the USA directed towards reactor pressure vessels indicates that low alloy steel weldments can be assessed using parent plate environmental fatigue crack growth data.[37] However, where coarse grained heat affected zone microstructures exist in thick section weldments then enhanced fatigue crack growth rates have been reported in high temperature water tests.[39]

When an assessment of fatigue crack growth is required for gas containers or pipelines then some thought has to be given to possible interaction between the gas and the crack tip. Nelson[40] has published results of a preliminary experimental survey of the fatigue crack growth behaviour of a plain carbon SAE 1020 pipeline steel exposed to various gaseous environments. Crack growth rates in carbon monoxide, carbon dioxide, oxygen, methane and natural gas are about the same as in air environment. The presence of water vapour caused some acceleration in growth rate but low pressure hydrogen caused an order of magnitude increase in growth rate at 1 Hz compared to air. Similar effects of hydrogen occur in higher strength structural steels, e.g. in HY-80 steel with a yield strength of 655 MPa low pressure hydrogen produced increased fatigue crack growth rates up to $30 \times$ faster than in air.[41]

Cathodic corrosion protection is used for certain pressurised components like pipelines but it may produce unexpectedly detrimental effects on fatigue crack growth rates. Very large accelerations in fatigue crack growth rate, up to $\times 50$ compared with air, have been observed by Vosikovsky[42] for C–Mn pipeline steel (yield strength 460 MPa) tested in salt water under applied cathodic protection. These large effects are attributed to the hydrogen released electrochemically close to the crack tip.

5. CONCLUDING REMARKS

It is now possible to predict with fair accuracy the life of a pressurised component subject to fatigue loading. Where environmental aspects are insignificant, fatigue crack growth data are available which can be used with confidence within given classes of materials. In these situations the major problems and uncertainties in practice are usually with aspects other than fatigue e.g. stress analysis or non-destructive testing. The three design or defect assessment analyses generally adopted are:

(i)   for a required cyclic duty, N, calculating the final crack size, a_f, to which a crack of known size at the beginning of life or at an inspection period, a_0, will grow,

(ii) for a known or assumed value of a_0, calculating the number of cycles necessary to grow the crack to the critical size for catastrophic fracture,

(iii) from a knowledge of the cyclic duty, N, and the critical crack size, a_f, calculating the crack size, a_0, which must be detectable by NDT before the plant enters service.

Several areas of uncertainty remain in the field of fatigue crack growth. Environmental effects are probably the most important because they can introduce lack of pessimism into analyses. Beneath K_{ISCC}, even notionally unaggressive environments like water can produce significant accelerations in fatigue crack growth rate. The actual increase depends on the detailed cycle in use e.g. its stress ratio and loading rate. For important assessments it may be necessary to test in conditions which simulate the actual cycle and environment as closely as possible. A second area of uncertainty involves understanding the behaviour close to the threshold for crack growth, ΔK_0. This applies particularly to cracking at regions of stress concentration. Just as with environmental effects, there is a dearth of data on weldments. The third area of uncertainty concerns the interaction between different types of fatigue cycle, i.e. spectrum loading or overload effects. Significant effort is currently being expended on this problem particularly via crack closure ideas. Overload effects are generally ignored in pressure vessel analyses and overload retardation of fatigue crack growth may significantly aid pessimism in such analyses.

REFERENCES

1. FROST, N. E, MARSH, K. J. and POOK, L. P. *Metal Fatigue*, 1974, Oxford University Press, Oxford.

2. FORSYTH, P. J. E. and RYDER, D. A. *Metallurgia*, 1961, **63**, 117.
3. LAIRD, C. and SMITH, G. C. *Phil. Mag.*, 1962, **7**, 847.
4. McMILLAN, J. C. and PELLOUX, R. M. N. 'Fatigue crack propagation', 1967, *ASTM STP* 415, Philadelphia.
5. MIYAZONO, S. and SHIBATA, K. 3rd International Conference on Structural Mechanics in Reactor Technology, London, Paper 94/8, 1975, Commission of the European Communities, Brussels.
6. PARIS, P. and ERDOGAN, F. *J. Basic Engng.*, *Trans. ASME*, 1963, **85**, 528.
7. BARSOM, J. M. *J. Eng. Ind.*, *Trans. ASME Series B*, 1971, **93**, 1190.
8. BARSOM, J. M., IMHOF, E. J. and ROLFE, S. T. *Engng. Fracture Mech.*, 1971, **2**, 301.
9. RICHARDS, C. E. and LINDLEY, T. C. *Engng. Fracture Mech.*, 1972, **4**, 951.
10. RICCARDELLA, P. C. and MAGER, T. R. 'Stress analysis and growth of cracks', 1972, *ASTM STP* 513, Philadelphia.
11. |MAGER, T. R., LANDES, J. D., MOON, D. M. and McLAUGHLIN, V. J. |Heavy Section Steel Technology Technical Report No. 35, 1973, Westinghouse Electric Corporation.
12. *An Assessment of the Integrity of PWR Pressure Vessels*, 1976, United Kingdom Atomic Energy Authority, HMSO, London.
13. CHELL, G. C. Central Electricity Research Laboratories, private communication.
14. ELBER, W. *Engng. Fracture Mech.*, 1970, **2**, 37.
15. VON EUW, E. F. J., HERTSBERG, R. W. and ROBERTS, R. 'Stress analysis and growth of cracks', 1972, *ASTM STP* 513, Philadelphia.
16. 'Fatigue crack growth under spectrum loads', 1976, *ASTM STP* 595, Philadelphia.
17. ARPA Materials Research Council meeting on 'Fatigue Crack Growth', 1976, La Jolla.
18. LINDLEY, T. C. and RICHARDS, C. E. *Mat. Sci. and Engng.*, 1974, **14**, 281.
19. BACHMANN, V. and MUNZ, D. 'Fatigue testing and design', Conference of the Society of Environmental Engineers Fatigue Group, Paper 35, Soc. Environ. Eng., 1976, London.
20. POOK, L. P. 'Stress analysis and growth of cracks', 1972, *ASTM STP* 513, Philadelphia.
21. WEISS, V. and LAL, D. N. *Metall. Trans.*, 1974, **5**, 1946.
22. LINDLEY, T. C. and RICHARDS, C. E., to be published.
23. JACK, A. R. and PATERSON, A. N. Inst. Mech. Engineers Conference on 'The Influence of Environment on Fatigue', 1977, London.
24. FROST, N. E., POOK, L. P. and DENTON, K. *Engng. Fracture Mech.*, 1971, **3**, 109. See also reference 1.
25. BROEK, D. Nat. Aerospace Inst. Amsterdam Report TR 72134, 1972. Quoted in BROEK, D., *Elementary Engineering Fracture Mechanics*, 1974, Noordhoff, Leyden, p. 320.
26. SMITH, R. A. and MILLER, K. J. *Int. J. Mech. Sci.*, 1977, **19**, 11. See also SMITH, R. A., *Fracture Mechanics in Engineering Practice*, ed. P. Stanley, Applied Science Publishers, 1977, London.
27. FORMAN, R. G., KEARNEY, V. E. and ENGLE, R. M. *J. Basic Engng.*, *Trans. ASME*, 1967, **89D**, 459.

28. HEALD, P. T., LINDLEY, T. C. and RICHARDS, C. E. *Mat. Sci. and Engng.*, 1972, **10**, 235.
29. IRVING, P. E. and McCARTNEY, L. N. *Fatigue 1977*, The Metals Society, 1977, Cambridge.
30. *Corrosion Fatigue: Chemistry, Mechanics and Microstructure*, eds. DEVEREUX, D. F., McEVILY, A. J. and STAEHLE, R. W. National Association of Corrosion Engineers, 1972, Houston.
31. 'The Influence of Environment on Fatigue', Inst. Mech. Engineers, 1977, London.
32. BARSOM, J. M. *Engng. Fracture Mech.*, 1971, **3**, 15.
33. BARSOM, J. M., 1972. In reference 30, p. 424.
34. ATKINSON, J. D. and LINDLEY, T. C. In reference 31, p. 65.
35. DAWSON, D. B. and PELLOUX, R. M. *Met. Trans.*, 1974, **5**, 723.
36. DOKER, H. and MUNZ, D. In reference 31, p. 123.
37. BAMFORD, W. H., MOON, D. M. and CESCHINI, L. J. Fifth Water Reactor Safety Information Meeting, Gaithersburg, Nov. 1977.
38. KONDO, T., KIKUYAMA, T., NAKAJIMA, H., SHINDO, M. and NAGASAKI, R. In reference 30, p. 539.
39. SUZUKI, M., TAKAHASHI, H., SHOJI, T., KONDO, T. and NAKAJIMA, H. In reference 31, p. 161.
40. NELSON, H. G., 1975, Conference on Effect of Hydrogen on Behaviour of Materials, AIME.
41. CLARK, W. G., 1974, 'Hydrogen in Metals', ASM.
42. VOSIKOVSKY, O. *J. Engng. Mat. & Tech., Trans. ASME(H)*, 1975, **97**, 298.
43. MOGFORD, I. L. and PRICE, A. T. 1972, 'Welding research related to power plant', 1972, CEGB.
44. BERNARD, P. J., LINDLEY, T. C. and RICHARDS, C. E. 1976. 'Fatigue crack growth under spectrum loads', 1976, *ASTM STP* 595, Philadelphia.

Chapter 8

PROBABILITY ASPECTS OF RELIABILITY ASSESSMENT

G. O. JOHNSTON

The Welding Institute, Research Laboratory, Abington, UK

SUMMARY

As practical advances are being made in welding technology, there is a growing demand for statistical analyses, especially in relation to the safety of nuclear pressure vessels. After discussing some of the background to the problem, the new probabilistic fracture mechanics approach is explained. The failure probability is synthesised into components, each of which may vary statistically. In a simple analysis, failure is given by the interaction of the initial and critical crack size distributions, which may then be changed by nondestructive examination (and subsequent repair of defects) or fatigue crack growth. Finally, possible future developments are considered, in particular, those pertaining to the accumulation of data, which are very scarce at present.

1. INTRODUCTION—ESTABLISHING THE REQUIRED RELIABILITY

How reliable is a structure? What is its probability of failure? How safe is it? These are questions which are being asked more and more frequently as the concern for safety, especially in relation to pressurised structures, increases. At a first glance, these may seem simple requirements deserving a concrete figure as an answer, to be given without further thought. In reality, the problems associated with answering these questions are quite complex and so a new field has developed, that of probabilistic fracture mechanics.

203

The dictionary definition of reliability is soundness and consistency of quality. A reliable component is one that can be depended upon with confidence. In high integrity structures such as pressurised components especially in nuclear power plants, reliability is particularly required when failure can cause much devastation both in human life and cost of equipment. For example, it appears that the reliability of nuclear reactors must be required to be considerably greater than that of even conventional pressure vessels. The requirements for nuclear pressure vessel integrity are discussed by O'Neil and Jordan[1] in terms of the permissible Iodine-131 release per year. For releases of Iodine-131 between about 10^2–10^8 Curies the safety criterion is taken as:

Release of Iodine-131 (in Curies) × frequency of release per year ≤ 1

Three main categories of pressure vessel failures are considered, namely:

A—catastrophic failure possibly leading to a breach in the vessel wall;
B—gross ductile rupture where the containment structure is not at risk though leakage may increase; and
C—small scale rupture or limited leakage within the design capacity of the vessel.

For category A failures, it is postulated that not more than about 10^7 curies will be released so that a (maximum) allowable frequency of events (releases) is 10^{-7} per vessel year according to the safety criterion. However, for complete release to the atmosphere, it is necessary to suppose that all engineered safeguards and containment fail completely. This seems unlikely to occur except perhaps around 1 in 10 to 1 in 100 cases, so that the required catastrophic failure rate would be 10^{-5} to 10^{-6} per vessel year.

In gross ductile rupture (B) complete failure, as described above, is unlikely and a decontamination factor of 10^{-2} is assumed giving an acceptable frequency of 10^{-5} per vessel year.

Small failures (under category C) are unlikely to release more than about 10^3 curies of I-131, allowing a frequency between 1 and 10^{-3}. However, since small leakages may increase, the acceptable frequency is assessed at about 10^{-3}–10^{-4} per vessel year.

This example is quoted in some detail as an illustration of the way in which a required failure probability may be established for any particular application.

Having set the standards, the next question to be considered in this context is whether they can realistically be met. To this end, O'Neil and Jordan first consider published statistics on nuclear pressure vessel integrity, amounting to only 2×10^3 reactor vessel years of experience, with no reported category A or B failures. Clearly such figures cannot be used to give any meaningful assurance. Considering data on conventional pressure vessels reported by Phillips and Warwick[2] (and discussed in more detail in the next section), there appears to be a difference of 10^{-1}–10^{-2} between the desired nuclear vessel failure rate and the corresponding rate for conventional pressure vessels. So the question still remains as to whether it could be demonstrated that a reactor pressure vessel could be made up to the required standard and to answer that an alternative approach is necessary, using the available data in an improved way.

Many factors contribute to reliability and these often change both systematically and at random. In the past, structures have been designed using a simple factor of safety approach to calculate a safe lifetime over which the applied loads will be sustained. Each variable in the equations governing the operation of the system (for example, fracture mechanics or fatigue S–N relationships) was given a reasonable value drawn from past results or informed judgement and a predicted life obtained. This was then divided by a factor of safety which reduced it to give a 'safe' maximum operating life and also gave some answer to the third initial question, 'how safe is a structure?' It is conceivable that this is unrealistically conservative, but it is the price paid for simplicity. However, the degree of conservatism (if any) given by such an approach is not known and in the design procedure there is a growing requirement to quantify the true safety of a structure. Guesses are no longer sufficient and even an undemonstrated degree of over-conservatism is not totally acceptable, so that there is a demand particularly from certain industries for a statistical justification behind the calculations.

One possible method of defining the probability of a structural failure is to use historical data from past failures and non-failures. This is a reasonably simple approach given a sufficient quantity of data, but its drawback is that only specific structures are considered. Extrapolation to other cases where no data may exist is probably suspect since a change in one of the key parameters may influence the results in an unpredictable way.

Alternatively, engineering models based on an understanding of the failure modes and statistical distributions of the controlling parameters can be developed and applied to calculate failure probabilities. Again, the problem is a lack of data. The distributions of the variables have often to be calculated from a small quantity of data, so that errors may be introduced because of poor assumptions (the 'correct' ones being as yet unknown).

Another disadvantage is the relative complexity of the approach, which requires more input parameters where errors may again occur.

Ideally, a combination of the two methods is necessary so that past experience and results may be supplemented by specially designed laboratory tests to give a clearer understanding of the variation of the parameters involved in the engineering model. Certainly, the results would, in a strict sense, only be valid in a particular case, but by isolating one variable, such as temperature, and keeping the rest constant as far as possible, predictions about altered or new systems may be feasible.

While the results and methods of analysis developed here are specifically for pressurised components, including nuclear experiences, the lessons learned are for the most part very much more general in their applicability to other engineering structures. Many of the problems, such as a lack in data or defects, are present in other situations and the concept of probabilistic fracture mechanics and fatigue crack growth has been widely adopted.

2. HISTORICAL BACKGROUND

Reliability is of particular importance when related to pressure vessels because of the disastrous results associated with a 'catastrophic' failure. A key variable in any analysis is that pertaining to defects of flaws in the structure under question and therefore several surveys have been conducted in order to quantify the defects present in pressure vessel seams. One of the earliest papers is by Phillips and Warwick.[2] They consider results for a five year period, but rely solely on the memories of people working with the vessels to gain information on vessels which failed prior to service or during operation. Though at first sight this may not seem a very accurate assessment of defects present in pressure vessel seams, it is likely to be reasonably reliable since the major events are fortunately comparatively rare and so are more likely to be remembered. The results are classified according to cause of failure, detection methods and causes of cracks, giving the percentage of each compared with the total number of failures. Failures are defined as potentially dangerous or catastrophic, occurring either during construction or service. The numerical indications of failure probabilities from the survey were found to be: for in-service failures, 10^{-3} and 10^{-4} per vessel year for potentially dangerous and catastrophic occurrences respectively, and: prior to service, 10^{-4} per vessel year in both cases. The most probable time for incidents appeared to be during the first half of the design life.

An extension to this work is given by Smith and Warwick.[3] Rather than relying on human memory, a survey similar to the previous one was carried out using evidence documented over the five years between the two reports. The results are very similar, though nondestructive examination methods and the extent of their application had clearly improved.

Several other papers, including the USAEC Advisory Committee on Reactor Safeguards (ACRS) Report on the Integrity of Reactor Vessels[4] and Bush[5] give data on the frequency of failures in pressure vessels. However, in comparing these, care is required since the exact definition of a failure (for example, catastrophic or disruptive) varies according to the author. This problem is illustrated in the first of these reports (Table 5.1) where failures are divided into three, namely non-critical (non-disruptive), potentially disruptive and disruptive. However, Phillips and Warwick[2] use only two groups, that is potentially dangerous and catastrophic with the split being somewhere in the potentially disruptive class. In the ACRS Report on the Integrity of Reactor Vessels,[4] failure probabilities of 10^{-5} and 10^{-6} per vessel year are calculated for a United States non-nuclear boiler pressure vessel and a reactor vessel respectively. However, there must always be uncertainties on the applicability of values derived from such surveys to other situations, involving vessels designed, fabricated and operated to differing requirements.

3. SYNTHESIS INTO COMPONENT PROBABILITIES

3.1. General Description
In the past, the safety of engineering structures has been assessed by considering component parts which each contribute a portion to the possibility of failure. The initial step was to determine which parameters were likely to influence the results and then determinate values were assigned to each. Statistical variations were ignored. A similar synthesis is required when taking such variations into consideration, though the important variables will no longer have a determinate value but will be described by a 'distribution' of values.

One of the first calculations of a failure probability for pressure vessels was by O'Neil and Jordan[1] using the data presented by Phillips and Warwick.[2] They used the deterministic approach described above, but their division of the failure probability works equally well in a more complex (statistical) description and so serves as an example of the important components in calculating, P_F, the failure probability.

Perhaps the most crucial consideration is the probability of failure in material, P_M, which may be caused, for example, by defects in a weld length. Closely linked with this are failure in design, P_D, due to crack growth and low toughness and the probability of errors in construction, P_C, such as heat treatment or fabrication.

Checking a vessel for flaws plays an important role in its operational life. These may be found by a pre-service pressure test, P_{PT}, or in service, stress wave acoustics, P_{AT}, ultrasonics, P_{US}, visual examination, P_{VE}, and leakage, P_L, may all be used and so must be included in the analysis. A final factor, P_X, which might be included would cover external modes of failure such as sabotage, but this was excluded from the calculation.

The assumed values for each of the variables are given in Table 1 and the overall failure probability

$$\begin{aligned}
P_F = &\ P_D \times P_{PT} \times P_{AT} \times P_{US} \times P_L \times P_{VE} \\
&+ P_M \times P_{PT} \times P_{AT} \times P_{US} \times P_L \times P_{VE} \\
&+ P_C \times P_{PT} \times P_{AT} \times P_{US} \times P_L \times P_{VE} \\
&+ P_X
\end{aligned}$$

was calculated to be 8×10^{-7} per vessel year.

In the statistical approach, flaw length and critical crack length are treated as variables rather than determinates and in a simple analysis the failure probability is given by the interaction of the two distributions. In

TABLE 1

SUMMARY OF ASSESSED INDIVIDUAL FAILURE PROBABILITIES[1]

Item	Assessed probability of failure per five-year cycle	Selected values for equation
P_D	5×10^{-5}	5×10^{-5}
P_M	5×10^{-5}	5×10^{-5}
P_C	1.5×10^{-4}	1.5×10^{-4}
P_{PT}	$1-0.985$	1
P_{AT}	$1-10^{-2}$ (first inspection) $10^{-1}-10^{-2}$ (repeat inspection) 10^{-2} (on-line inspection)	0.5 for design defects 10^{-1} for material and construction defects
P_{US}	$1-10^{-1}$ (first inspection) $10^{-1}-10^{-2}$ (repeat inspection)	0.5 for design and material defects 10^{-1} for construction defects
P_V	$1-10^{-1}$	0.5
P_L	$1-10^{-1}$	0.5
P_X		0.0

practice, crack growth occurs and in the analysis some growth law must be assumed. Often the empirical (deterministic) Paris law

$$\frac{da}{dN} = C(\Delta K)^m$$

is used, but really, C and m should be treated as variables. So, too, must the description of the reliability of nondestructive examination methods allow for random variations. This provides several varying parameters to put into the analysis and at present any others are likely to be given deterministic values in order to keep the calculations as simple as possible.

An assessment of the general problem including references to current work and consideration of the various parameters, such as defect size, fracture toughness and residual stress is given by Nichols.[6] In particular, the probability density function (statistical) approach has been developed for situations where the major risk is believed to be that of fast fracture. Any non-negative function, f(x), which has unit area under the curve is called a probability density function (pdf), and its characteristic property is that the proportion of observations between $x = a$ and $x = b$ is the area under the curve from a to b. This is consistent with unit total area since it implies that all observations lie between $x = -\infty$ and $x = +\infty$. Some commonly used distributions are shown in Fig. 1.

Early work in this field includes that by Lomacky et al.[7] who first considered the fatigue crack growth of a single crack in a spherical pressure vessel shell. Each load cycle produces a new crack size distribution which is calculated. The analysis is then extended to take account of the number of cracks occurring throughout a weld length assuming it is a random variable with a Poisson probability (see Fig. 1(a)). Jouris and Shaffer[8] quote data which gave a three-parameter Γ distribution (Fig. 1(h)) for K_{Ic} with temperature dependent parameters, though their analysis is somewhat simplified in that they have assumed that K_I is not a random variable.

As a specific example of pressure vessel reliability, Arnold[9] calculates the probability of failure as a function of general membrane stress, burst pressure and cyclic stress amplitude. He applies the method to typical induced stress and material strength distributions, which he takes as normal (Fig. 1(d)), based on available data. The induced stress distribution closely approximates to the normal distribution where there is a high incidence of peak pressures (which are assumed to occur as a factor in fatigue analysis) on either side of the design pressure. The theory is then extended to allow for cyclic stress distributions and for simplicity the Weibull distribution function (Fig. 1(g)) is used to generalise the distribution of allowable stress

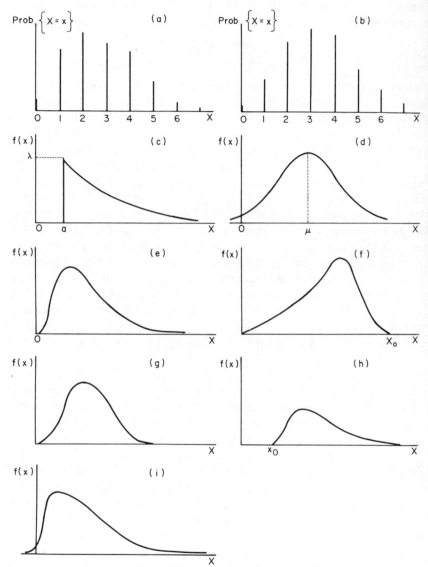

FIG. 1.   Discrete probability distributions (Prob $\{X = x\}$) and continuous density functions f(x). (a) Poisson, (b) binomial, (c) exponential, (d) normal (Gaussian), (e) lognormal, (f) negative lognormal, (g) Weibull, (h) three-parameter gamma, (i) extreme value, Type 1 of maxima.

for all numbers of cycles and all stress levels. After this follows a numerical example giving the failure probability as 10^{-6} per vessel year. In conclusion, it is suggested that consideration be given to lowering the allowable cyclic stress amplitude and that allowable membrane stress intensities may be increased without producing a significant change in failure probability.

3.1.1. *Probabilistic Fracture Mechanics Approach*

Assuming the pre-service flaw size distribution is known (whether this be the as-fabricated defect distribution or a value obtained by modifying this as-fabricated value by an allowance to cover the detection of flaws by the appropriate nondestructive examination method), with probability density function $f(x)$, then the failure probability of a structure is given by the interaction of an actual flaw length and a critical crack length, as described by Nichols.[10]

Consider first of all flaws with a length between x and $x + dx$; then the probability of interaction of one of these flaws is given by the probability that it is larger than the critical flaw size, a_{crit}, for the material and the conditions under consideration. Now multiply this by the probability of having a flaw of size x to $x + dx$, which is $\text{Prob}\{x \le \text{flaw length} \le x + dx\} \approx f(x)\,dx$ for a small interval dx. Thus for these flaws the interaction probability dP is

$$dP = \text{Prob}\{a_{crit} \le \text{flaw length}\} \cdot \text{Prob}\{x \le \text{flaw length} \le x + dx\}$$

where

$$\text{Prob}\{a_{crit} \le \text{flaw length}\} = \int_0^{\text{flaw}} g(a_{crit})\,da_{crit}$$

assuming $g(a_{crit})$ is the probability density function for the critical crack size distribution. Therefore,

$$dP = \left\{\int_0^x g(a_{crit})\,da_{crit}\right\} f(x)\,dx$$

Summing over all possible flaw lengths x gives the total interaction probability, P, as

$$P = \int_0^\infty \int_0^x g(a_{crit})\,da_{crit}\,f(x)\,dx$$

Figure 2 illustrates the above calculation graphically. For the flaw distribution, the probability density function $f(x)$ is plotted. For the critical defect distribution, the 'cumulative' distribution function is shown, that is $\int_0^x g(a_{crit})\,da_{crit}$ for varying x from zero to infinity. Since a probability density function has unit area under the curve by definition, the graph of $\int_0^x g(a_{crit})\,da_{crit}$ tends to unity as x tends to infinity. Then the integrand $\int_0^x g(a_{crit})\,da_{crit}\,f(x)$ for P is just formed by multiplying corresponding values

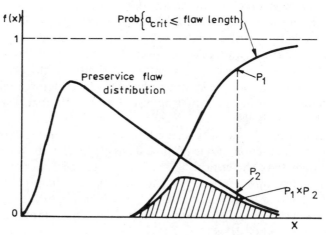

FIG. 2. Interaction (given by shaded area) of flaw and critical crack distributions.

on the two curves at a particular value of x, e.g. P_1 and P_2 on Fig. 2. This gives the curve bounding the shaded area which is itself the required interaction probability, P.

3.2. Individual Components

3.2.1. Flaw Distribution

The as-fabricated flaw size distribution is a key input to any analysis and, unfortunately, probably least is known about it. Closely linked with this is the role played by nondestructive examination in detecting defects, which may or may not be repaired. Thus, if the initial flaw size distribution and the probability of detection of a defect are known, then the distribution of flaws when the structure is put into service can be deduced.

One of the greatest problems in this field seems to be the assessment of the as-fabricated distribution of crack like defects. Nichols[11] states that 'assurance of weldment reliability implies effective control at all stages of design, fabrication and operation in service' so that materials and processes should be chosen to minimise the production of flaws. Despite this care, there are still some flaws present in any physical structure. For the most part, there are few data available to give even a starting point. Most of the estimates seem to be based on experience and human judgement, though this can give quite a good indication of the order of magnitude of failure probabilities as shown earlier by comparing the results of the surveys by Phillips and Warwick[2] and Smith and Warwick.[3]

It is reasonable to expect that there will be a large number of small defects with decreasing number of larger ones, so that there is a high probability of obtaining defects between zero and some small flaw size. Hence the exponential distribution (Fig. 1(c)) seems a viable model, though some authors prefer to use the lognormal distribution (Fig. 1(e)).

Although some data are available on the distribution of defects as detailed earlier, most are in the form of statements such as 'a weld length contains a certain percentage of defective material' or a value is quoted for the expected number of cracks greater than some given length in a weldment. Little is known about the actual size, location or orientation of the defects particularly for the larger defects which so affect the end result. Furthermore, the defects recorded are only those found by nondestructive examination, (usually radiography and ultrasonics in practice) and so the data reflect also, the reliability of the examination method in use.

Salter and Gethin[12] conducted a survey of the type and frequency of defects found by radiography in the main seams of pressure vessels. The significance of these defects was assessed together with an estimate of the number which might be unacceptable, by assuming all planar defects, such as cracks or lack of fusion, are not acceptable whereas porosity and slag inclusions are allowable. For each defect an estimate was also made of the volume of weld metal removed and repaired.

It was found that 81 % of all defects could be classed as non-critical in the terms above and that 87 % of the weld metal volume resulted from these. Therefore, repair of only the planar defects would have represented a considerable saving in repair time and weld metal cost. Of the 599 seams examined, only 27 % contained critical defects.

Generally, pressure vessel manufacturers will record just sufficient information in order that unacceptable defects can be repaired. With the modern computerised systems a balance must be sought between the high cost of storing information and the obvious desirability of keeping a record of everything. In practice, the cost factor 'wins' so that a minimum of information is stored and the system is only enlarged when there is a need for further data, though often this is too late to solve the particular problem in hand. Unfortunately, this means that some data which might be useful in a probabilistic analysis are discarded at present, for example data on small acceptable defects may be rejected as being unnecessary information for repairing a weld length. Turning to the nondestructive examination aspect, some test results may be biased by prior knowledge when two different NDE techniques are stipulated in the specification. Separation of these could perhaps allow a comparison between the flaw detection capabilities of radiography and ultrasonics which are the most common methods in use. In this case, care is required when analysing the results since a specification of 10 % X-ray and 100 % ultrasonics may apply to just part of the weld, so that some of the weldment does not require to be tested.

3.2.2. Nondestructive Examination

Ideally, the distribution of as-fabricated defects should be truncated when combined with the probability of detection function, so that all defects of a certain size (and larger) are always found, as described by Volchenko.[13] Tang[14] compares actual (a_a) and measured (a_m) crack sizes by passing the flaw size distribution through a filter defined by the 'detectability function'. He obtains an expression of the form:

$$a_a = C_0 + C_1 a_m + C_2$$

where C_0 and C_1 are regression coefficients and C_2 is a calibration error assumed to be normally distributed with zero mean. Different non-destructive testing methods, of course, filter the flaw size distribution by different amounts. A general assessment of the problem is given by Hansen[15] and he also emphasises the uncertainty involved when an inspector has to decide whether or not to reject a particular defect. If he encounters a defect with a true maximum size equal to the critical size, then there is a 50 % chance of rejection, since in some cases he will reject the defect and in others he will judge it to be just acceptable. This problem is dealt with in more depth by Johnson.[16]

Returning to the actual problem of detection reliability, one way to combat the lack of data on this aspect is to perform experimental work specially designed to try to quantify the reliability of the various available nondestructive examination techniques. In theory, this sounds an easy job, but in practice there are several complications. One of the major ones is the human element which can have quite a significant influence on the results when assessing whether flaws are really present or what size they may be since some operators are more experienced than others. Even the production of test pieces may not be straightforward, since if the welder knows the purpose of the experiment, then psychologically he will tend to produce a better weld and consequently fewer defects. Necessarily the test piece will also be a simplification compared with geometries encountered in practice. Some experimental work has been done, but mostly in relation to aluminium; however, the methods would seem to be viable for other materials. The results are given here only as an indication of what may be found, but it must be emphasised that their applicability to other materials is highly suspect in the absence of data to the contrary.

The work by Packman et al.[17] was on tubes of vanadium modified steel and 7079-T6 511 aluminium alloy in constant amplitude fatigue tests. They define an assurance index as a cumulative measurement of the sensitivity, accuracy and precision of each nondestructive examination method, where the sensitivity is the proportion of flaws found by that particular technique, the accuracy is that of the flaw size estimation and the precision is based on the standard deviation of the measurements involved. No cracks smaller than 2·5 mm were detected consistently with any nondestructive examination method, but all gave high accuracies with longer cracks. The sensitivity of the shear wave ultrasonic technique was superior to all the other methods examined, and magnetic particles were better than penetrants for detecting and measuring cracks in the steel. In all cases, X-rays did not give very good results. The maximum assurance index obtained was 90% showing the need for further improvements in techniques.

Further work by Packman, Klima et al.[18] actually tries to assess the reliability of nondestructive examination by assuming a binomial distribution (Fig. 1(b)) for finding or missing a flaw. They also suggest a test programme (including the choice of sample size) for obtaining detection probabilities. Briefly, a number of specimens are fabricated with 'known' defects as far as is possible and at least an equal number are made with 'no' defects. The specimens are then mixed up so that the inspector does not know the flawed specimens from the rest. Then the situation most nearly represents real life conditions, where flaws may or may not exist. Obviously, it takes a lot of skill to put a specific defect of a particular size at a given location in a weld, but after testing the specimens may be broken up to check on the actual flaws present, and a direct comparison with the detected flaws can be made. Typical flaw detection data are presented in Fig. 3 for comparison of various inspection procedures and data analysis methods that are detailed in reference 18.

Similarly, Rummel et al.[19] performed experiments on aluminium alloy using ultrasonic, eddy current, penetrant, holographic interferometry, X-ray and acoustic emission techniques. They concluded that the first four

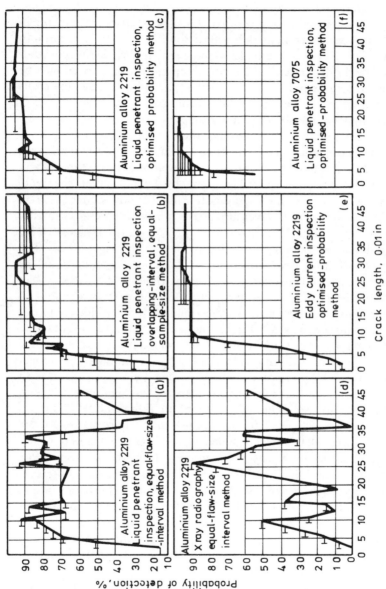

FIG. 3. Charts of lower bound probability of detection of fatigue cracks of various lengths in specimens made of two different alloys and inspected by three different procedures. Three different methods were used to plot the inspection data. *Caution:* these charts were derived from inspection data for specific inspection procedures used on a specific specimen design. The charts are not recommended for use as a basis for designing other structures nor for certifying other inspection procedures.

were best for small tight cracks and that ultrasonics produced least scatter of data. The X-ray method was again least reliable for crack detection so that these results seem compatible with those by Packman, Klima *et al.*, above.

More specifically, in ultrasonic testing, the work done by the PVRC Study Group and reported by Buchanan,[20] investigated two different procedures for examining a welded plate specimen containing fifteen 'known' defects. A number of different company teams examined the specimen independently and they had no prior knowledge about the defects. The first (older) method of testing had proved unacceptable in the past, because of the variation in results obtained by different users, so one objective was to see if the second (newer) method gave an improvement. The requirements of the new procedure are more exacting and effectively limited the number of variables associated with the ultrasonic testing, both with regard to equipment calibration and specimen examination requirements, for example:

(a)　The new procedure specifies glycerine as a transducer couplant whereas no particular one was previously specified.

(b)　A $1 \times \frac{1}{2}$ in lead metaniobate transducer must be used in the new procedure, whereas a 1×1 in quartz transducer was recommended (allowing other possibilities) in the old procedure.

The effect of cladding the specimen surface was also investigated. Assuming the defects were located exactly where they were intended, since the specimen had still to be metallurgically sectioned when the work was reported, the new procedure did give an improvement over the old, and also yielded less scatter between the testing teams. However, the effect of cladding, on the reliability of the new procedure appeared to be relatively small. In all cases, the numbers of defects correctly reported were rather low. An extension to this collaborative exercise is now being undertaken in Europe, and other similar experiments are in hand involving other NDE techniques. A final word of warning in this field is given by Davidson.[21] It is that in practice an added complication is that some area or volume of material may be inaccessible either from geometry, external factors or internal defects, and so must go uninspected.

3.2.3. *Fatigue Crack Growth*
Fatigue failure may be defined as failure due to repeated application of stress. The lifetime of a reactor vessel could conceivably be limited by the slow extension of sub-critical defects to critical dimensions under operating

conditions. Calculation of the rate of growth of postulated cracks in the vessel requires a knowledge of the basic laws of crack growth and of the imposed loadings resulting from the various operational transients. For a crack of depth a subjected to a stress range ΔK the rate of crack growth is usually expressed by the empirical Paris law

$$\frac{da}{dN} = C(\Delta K)^m$$

Up to the present time, the methods used to treat fatigue crack growth data have all been deterministic assuming Paris' equation holds with C and m having some determinate value. In the probabilistic concept nothing is necessarily sacred and it seems reasonable to suppose that even if the Paris law is accepted then C and m are likely to be subject to random (statistical) variations. For design purposes it is difficult to overcome the temptation to use a crack growth curve which forms an upper bound on the data available for the material of interest, but this may be over conservative. It may be that a graph of log da/dN against log ΔK is not a straight line but a curve. Indeed, Bamford, Shaffer and Jouris[22] consider this problem and refer back to some of their earlier work where it seemed that crack growth behaviour for reactor pressure vessel steels exposed to the reactor water environment did not follow the Paris law. Their more recent work develops a procedure for obtaining a global confidence limit on the crack growth rate as a function of stress range, so that the mean crack growth rate falls in this region. Tolerance limits for the crack growth rates are also estimated so that a large proportion of the data is expected to fall within these limits. The analysis can then be extended to test the quality of the purported model.

Having decided upon a particular crack growth law, this can then be used in the probabilistic analysis to calculate how the crack size distribution alters with the number of (applied) cycles. This can have a profound effect on the failure probability since the larger cracks tend to grow at a faster rate and this is the important part of the distribution which overlaps the critical crack size distribution. This effect is shown in Fig. 4.

Linked with this is the effect of in-service inspection, which can help to increase the lifetime of a structure. Assuming all the significant defects found in a pre-service inspection are repaired then by virtue of the uncertainties present in the reliability of any flaw detection method, there will still be some flaws present in the structure. After a time in service these will grow according to some fatigue crack growth law. In-service inspection will then modify the distribution assuming all significant defects that are found are repaired and hence the lifetime will be increased, even though that

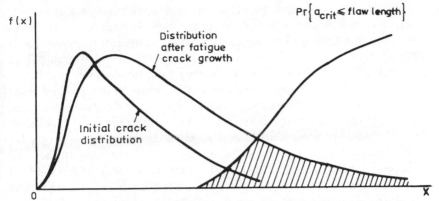

FIG. 4.   Change in flaw distribution because of fatigue crack growth.

inspection may itself be imperfect. This process can be repeated at fixed intervals of time as long as is required. Alternatively, rather than specifying a constant inspection interval, it is possible, and may be preferable, certainly for high integrity structures, to calculate a variable interval which gives a maximum permitted probability of failure. A mathematical treatment of fatigue crack growth and in-service inspection is given by Harris.[23]

3.2.4. *Material Toughness*

Again there is a problem in lack of data in particular for actual measurements of relevant fracture toughness properties. Few (if any) such data exist in sufficient quantity to permit a statistical analysis. Parameters such as Charpy V energy, K_{Ic} or crack opening displacement (COD) can be studied to give some idea of the critical crack size distribution. One possible approach is to collate all available data (with varying conditions) and group them together in some way to give an idea of the distributions involved.

Such an approach is taken by Wells[24] who uses a world-wide collection of results on Charpy V energy, obtained by Lloyds Register of Shipping, in order to derive a distribution of critical crack lengths. The normal distribution seems to fit these data, and Wells then combines this with a half normal distribution representing the flaw size, to obtain an expression for the interaction probability, P. In this case, P is found to depend heavily upon a factor which is inversely proportional to the root mean scatter. Of course, one would expect much more scatter in this case caused by the variety of conditions put together and it could be that this would mask the true distribution shapes.

However, things are perhaps not so serious as for the defect distribution, since it is possible to perform relatively easy laboratory tests under controlled conditions (for example, constant temperature, same plate, welding process, operator, etc.). However, care is then required in interpreting the results in the light of the assumptions made.

In probabilistic analyses, some distribution must be assumed in order that it can be combined with the defect distribution as described earlier. In many instances in the literature either the lognormal (Fig. 1(e)) or Weibull (Fig. 1(g)) distributions seem to be chosen to represent K_{Ic}.

This probabilistic fracture mechanics approach is taken by Becher and Pedersen[25] who evaluate the failure probability for a reactor vessel assuming linear elastic fracture mechanics. Both K_I and K_{Ic} are treated as statistical variables depending on material, location and time. The calculations are numerical and based on Monte Carlo methods so cannot be applied to new situations without access to the computer program.

More recently, a study group under the chairmanship of Dr Marshall[26] has investigated the reliability of pressure vessels in some detail. The normal or Weibull distributions are suggested for K_{Ic}, the exponential for the as-fabricated size of defect distribution and a slight variant on the exponential of the form $\varepsilon + (1 - \varepsilon) \exp(-\mu x)$ for the probability that defects of a given size are missed in an ultrasonic examination. In all cases available data are scarce. The conclusions again give the probability of failure as 10^{-6} per vessel year at the start of life and for a significant fraction of the vessel life thereafter, though the exact proportion will of course depend on the parameter assumptions.

This work and subsequent extensions to it are reported in more detail by Lidiard and Williams[27] where the failure probability of pressure vessels is estimated for normal, upset and test conditions (Q_f^N) and for emergency and fault conditions (Q_f^ε). A member of the vessel population is considered to fail if it contains a crack whose size is greater than the critical size for the stresses obtaining at that location and time. Despite the necessary simplifications and the uncertainties in the various parameters, the values calculated for Q_f^N seemed to be consistent with general experience on non-nuclear vessels and with present engineering judgement as to what that implies for nuclear vessels. The rates of fatigue crack growth rate are important for evaluating Q_f^N. However, the uncertainties in critical crack size for the emergency and fault conditions make it more difficult to obtain meaningful results for Q_f^ε, though it would appear that $Q_f^\varepsilon \sim Q_f^N$ for values between 10^{-4} and 10^{-2} per vessel year.

The main value in the calculations lies in the general pattern which emerges when some of the assumptions are changed rather than in the precise magnitudes calculated for the failure probabilities. In particular, the importance of information on crack growth rates is emphasised and an accurate knowledge of critical crack sizes is found to be of considerable value. Indeed, it would seem that a great deal of what one needs to know for the prediction of failure probabilities is contained in these parameters.

The effect of pre-operational pressure tests is also mentioned since they are almost always the most severe transients encountered in normal, upset and test conditions. Therefore, weak vessels are eliminated before they enter service and the statistical strength characteristics of those remaining will be altered since the distribution will (after testing) be truncated at the pressure test level. This truncation eliminates precisely the portion of the strength distribution, which is least known and has the greatest interaction with the applied load distribution. If there were no time variations, then these conditions would produce a zero probability of failure. Thus, the failure probability depends to a large extent on how the crack distribution changes with time, as described earlier.

4.   MATHEMATICAL TECHNIQUES

Given any set of data, the question arises as to what to do with it. Some possible treatments have been outlined briefly, for example, that of Phillips and Warwick.[2] However, for a probabilistic analysis, it is necessary to be able to fit a given distribution to the data. Which distribution to use may be decided from physical considerations; for example, flaw lengths are always positive, so a distribution allowing negative values is unrealistic, or alternatively where sufficient data exist, a statistical test might be carried out to find the distribution which best fits the data.

Assuming this decision has been taken, there are several methods which may be used to fit a given distribution to a quantity of data. Two of the most common are the method of moments and maximum likelihood estimation, the former usually being the simpler and consequently less accurate. Considering, for example, two parameter distributions, the method of moments involves simply calculating the mean and variance of the sample data and equating these equal to those of the assumed distribution. These two equations can in principle be solved for the two parameters. For the maximum likelihood estimation, a function L is defined and then maximised with respect to the two parameters. For a random sample of n

observations $x_1, x_2 \ldots x_n$ with probability density function $f(x)$ having parameters α and β then

$$L(x_1, x_2 \ldots x_n; \alpha, \beta) = \prod_{i=1}^{n} f(x_i)$$

Taking logarithms, differentiating with respect to α and β in turn and equating these to zero gives two equations to be solved for α and β. Often there is no analytical solution and numerical methods are required. In general, numerical integration is also required in the calculation of the failure probability.

Another technique which has received some attention recently in the calculation of failure probabilities is that of Monte Carlo simulation. This is perhaps best explained with reference to the initial and critical crack size distributions, though its real use is in more complicated calculations where several parameters are involved. Distributions are required for the two variables and these are fed into a Monte Carlo computer program. The process consists of a series of 'trials', each of which is either a failure or a success. In one 'trial', a value is taken at random from each of the initial and critical crack size distributions and the two compared. A failure occurs if the initial crack value is greater than the critical crack value, otherwise the trial is a 'success'. The failure probability is then simply the ratio of the number of failures to the total number of trials. The main drawback to the application of the method is that when considering probabilities of failure of the order of 10^{-6} (as in the case of pressure vessels) a large number of trials is required and this is expensive both in time and cost, though there are some techniques available for reducing this.

5. FUTURE DEVELOPMENTS

Probabilistic fracture mechanics is a relatively new field with a large scope for development in several areas. Obviously, the hardest, but most interesting requirement is to obtain information regarding the as-fabricated flaw size distribution. The ideal is to look at pressure vessels or other components which have been made to some specification and were expected to go into service but were scrapped because of some change in requirements. This could then be combined with efforts to investigate the reliability of various nondestructive examination techniques, the advantage being that the flaws present would be unknown until the vessel was cut up and so the results would not be biassed by prior knowledge. An alternative approach is to design special laboratory tests as suggested by the work of Packman, Klima et al.[18] described earlier.

Slightly easier to obtain are data on the strength distribution though any experimental tests will necessarily be on a particular material under specific conditions, and so care will be required in interpreting the results. As described earlier, the form of the fatigue crack growth law is also important in the analysis of pressure vessel reliability. Even assuming the Paris law as a first approximation, the variations of the 'constants' C and m need consideration.

Finally, an alternative approach, in the absence of concrete data, is to perform a parameter study, whereby the assumptions are varied to see what differences, if any, may result in the corresponding calculated failure probabilities. It is in this field that Monte Carlo techniques can be especially useful since they require a minimum of experimental data as a basis. This can really be quite a valuable part of the work since it should give an indication of which variables most affect the failure of a structure. Having isolated these, concentration can then be focussed on improving these in preference to other less important factors.

The concept of probabilistic fracture mechanics has developed in answer to a growing demand for a statistically based analysis of failure probabilities. There is an increasing awareness of the problems associated with the safety of nuclear pressure vessels, but improvements incur much greater costs so that fabricators want to know in which areas to concentrate their efforts. It is hoped that this new approach to the subject of pressure vessel failure will go some way to meeting this demand.

REFERENCES

1. O'NEIL, R. and JORDAN, G. M. 'Safety and reliability for periodic inspection of pressure vessels'. Proc. Conf. on Periodic Inspection of Pressure Vessels, 1972, I.Mech.E., London.
2. PHILLIPS, C. A. G. and WARWICK, R. G. 'A survey of defects in pressure vessels'. UKAEA Report AHSB(S)R162, 1968, HMSO, London.
3. SMITH, T. A. and WARWICK, R. G. Int. J. of Pressure Vessels and Piping, 1974, 2, 283.
4. Report on the Integrity of Reactor Vessels for Light-Water Power Reactors. USAEC Advisory Committee on Reactor Safeguards, Report WASH-1285, 1974, Washington.
5. BUSH, S. H. Trans. ASME, J. Pressure Vessel Technology, 1975, 97J, 54.
6. NICHOLS, R. W. Welding in the World, 1976, 14, 152.

7. LOMACKY, O., ANG, A. H.-S. and AMIN, M. ASME Paper 71-PVP-47, 1971.
8. JOURIS, G. M. and SHAFFER, D. H. Scientific Paper 75-IC51-FROBY-PI, 1975, Westinghouse Research Laboratories.
9. ARNOLD, H. G. ORNL-TM-3858, 1972.
10. NICHOLS, R. W., *Proc. 2nd. Int. Conf. on Pressure Vessel Technology*, 1974, III, 3, San Antonio, Texas.
11. NICHOLS, R. W. *Weld. J.*, 1975, **54,** 417s.
12. SALTER, G. R. and GETHIN, J. W. *Metal Fabrication*, 1973, **41,** 127.
13. VOLCHENKO, V. N. *Auto. Weld.*, 1975, **28,** 28.
14. TANG, W. H. *J. Test. and Evaluation*, 1973, **1,** 459.
15. HANSEN, B. 8th World Conf. on NDT, 1976, Paper 2A4, Cannes.
16. JOHNSON, D. P. *Materials Evaluation*, 1976, **34,** 121 and 136.
17. PACKMAN, P. F., PEARSON, H. S., OWENS, J. S. and YOUNG, G. *J. of Materials*, 1969, **4,** 666.
18. PACKMAN, P. F., KLIMA, S. J. *et al. Metals Handbook* 11, ASM, 1976, 414.
19. RUMMEL, W. D., TODD, P. H., FRECSKA, S. A. and RATHKE, R. A. NASA CR-2369.
20. BUCHANAN, R. A. Welding Res. Council Bull. No. 221, 18.
21. DAVIDSON, J. R. NASA TM-X-71969, 1973.
22. BAMFORD, W. H., SHAFFER, D. H. and JOURIS, G. M. 3rd. Int. Conf. on Pressure Vessel Technology, ASME, 1977, 815.
23. HARRIS, D. O. *Materials Evaluation*, 1977, **35,** 57.
24. WELLS, A. A. 'Design implications for materials selection'. Rosenhain Centenary Conf., 1975, The Royal Society.
25. BECHER, P. E. and PEDERSEN, A. *Nucl. Eng. and Design*, 1974, **27,** 413.
26. MARSHALL, W., UKAEA Report on the integrity of PWR Pressure Vessels, 1976.
27. LIDIARD, A. B. and WILLIAMS, M. *J. Brit. Nucl. Energy Soc.*, 1977, **16,** 207.

INDEX